Fourth Edition

How to Study Science

....

Fred Drewes

Professor Emeritus, Suffolk County Community College

Kristen L.D. Milligan, Ph.D.

Boston   Burr Ridge, IL   Dubuque, IA   Madison, WI   New York   San Francisco   St. Louis
Bangkok   Bogotá   Caracas   Kuala Lumpur   Lisbon   London   Madrid   Mexico City
Milan   Montreal   New Delhi   Santiago   Seoul   Singapore   Sydney   Taipei   Toronto

We dedicate this book to the students who continue to take responsibility to improve their study skills to attain their collegiate goals.

—Fred Drewes & Kristen Milligan

McGraw-Hill Higher Education

A Division of The **McGraw-Hill** *Companies*

HOW TO STUDY SCIENCE, FOURTH EDITION

Published by McGraw-Hill, a business unit of The McGraw-Hill Companies, Inc., 1221 Avenue of the Americas, New York, NY 10020. Copyright © 2003, 2000, 1997 by The McGraw-Hill Companies, Inc. All rights reserved. No part of this publication may be reproduced or distributed in any form or by any means, or stored in a database or retrieval system, without the prior written consent of The McGraw-Hill Companies, Inc., including, but not limited to, in any network or other electronic storage or transmission, or broadcast for distance learning.

Some ancillaries, including electronic and print components, may not be available to customers outside the United States.

This book is printed on recycled, acid-free paper containing 10% postconsumer waste.

10 11 12 13 14 QPD/QPD 09

ISBN: 978-0-07-234693-0
MHID: 0-07-234693-0

Publisher: *Martin J. Lange*
Senior sponsoring editor: *Patrick E. Reidy*
Developmental editor: *Anne L. Melde*
Marketing manager: *Tamara Maury*
Senior project manager: *Mary E. Powers*
Senior production supervisor: *Sandy Ludovissy*
Coordinator of freelance design: *Rick D. Noel*
Cover designer: *Mary K. Sailer, Spring Valley Studio*
Cover image: *©Getty Images, Female hiker standing on rock by Lori Adamski Peek, image no. eb2803-011*
Senior photo research coordinator: *Lori Hancock*
Executive producer: *Linda Meehan Avenarius*
Compositor: *Shepherd, Inc.*
Typeface: *10/12 Times Roman*
Printer: *Quebecor World Dubuque, IA*

www.mhhe.com

CONTENTS

PREFACE

To learn science, you must first *want* to learn it. A teacher can lead you to the subject matter, but cannot make you learn! Successful students have learned how to learn.

If you are uneasy about learning science and have poor study skills, *How to Study Science,* fourth edition, will help you. However, you must incorporate good study skills into your academic routine and behavior. These skills must become second nature to you; you shouldn't have to think about learning to learn. The guide's helpfulness will depend upon your attitude, academic background, and present level of study skills.

Chapters 1 and 2 discuss the study of science and the natures of science courses and college instructors. Becoming familiar with the nature and structure of science courses will help you adapt to the challenge of these courses. This awareness will enable you to control your academic behavior.

Chapter 3 talks about the process of learning and introduces study skills to be used in and out of class. The checklist in this chapter provides a handy inventory of things you can do to help yourself learn.

Chapters 4 and 5 give you guidance to in-class skills. Self-discipline and self-direction are important parts of your study. You must also have good listening skills. In addition, you must be able to take notes and convert them into a learning tool.

Chapters 6 through 12 discuss time management, study sessions, use of a textbook, terms and symbols, figures and their uses, and assignments and reports.

Chapters 13 through 15 discuss writing essays, solving math problems, taking tests, and analyzing results. Written work you complete will be a measure of your ability to apply information you have learned. Chapter 15 suggests that you analyze results to identify the types of errors you make. This information should help guide your adoption and development of new study skills that will improve your learning and retention of information. Suggestions on the use of computers and computer software are included in a number of chapters.

Each chapter has a written exercise to emphasize and reinforce the information and study skills discussed in the chapter. The guide's helpfulness will be increased if you complete these exercises. An abbreviated answer guide is located in Appendix B to give examples of answers, but not complete answers. Compare your answers to those example answers. Some exercises do not have example answers. You should review your answers with your instructor and other students. In a sense, the saying "No pain, no gain" is true.

This guide can be used in several ways. Students may use this book to enhance their study skills, or instructors may use it as a text or supplement for orientation classes, college seminars, or science courses. We believe the whole book should be reviewed prior to or within the first few weeks of the semester. Specific chapters should be referred to later as a need arises. For instance, early in the semester you should know that the book contains information about lab reports, graphs, and writing essays. Later, as you get involved in doing these things, you can refer to specific content in the book that will help you complete the task. For example, it is important to know about preparation for tests from the beginning, but specific test-taking skills might be studied the week before the test.

Teachers might also use this book as a guide to help improve their teaching skills. For instance, teachers could present figures modeling the system of figure analysis (chapter 10) in their lectures rather than just describing selected parts of a figure.

This book could be used in orientation courses, as a supplement to a science course, or as a textbook for workshop minicourses before even taking a science course.

We would like to thank Robert S. Boyd, Auburn University; Gloria Early Payne, Saint Augustine's College; Ann E. Rushing, Baylor University; Stephen W. Wilson, Central Missouri State University; Caroline Peters, Spoon River College; Wendy E. Sera, Baylor University for critical review. We also thank Sandra Drewes, Director of Kellog Institute, Appalachian State University, Boone, N.C. for advice and comments on teaching and learning styles.

We hope you will find this book helpful. We would appreciate any feedback you care to give. Thanks.

Fred Drewes
Kristen Milligan

CHAPTER 1

The Study of Science

Objectives

When you have read this chapter, you will be able to answer these questions:

1. What is science?
2. How do scientists use the scientific method?
3. What is the connection between science and technology?
4. Why should I study science?
5. Am I ready to study science?
6. What does my attitude have to do with studying science?
7. What is the reward for studying science?

The sun shines. Clouds appear, and rain falls. Water soaks into the ground or runs off into rivers. Rock is eroded, and sediments are moved. Plants absorb water, carbon dioxide, and light. Animals consume the plants and scurry about claiming territory. All organisms struggle to survive.

What causes the sun to shine? Why does rain fall? What is water? Rock? A plant? An animal? A human being? Where did they all come from? Why do things work the way they do? One question leads to another; that is the nature of science.

What Is Science?

Science is the discipline that studies the **quantitative** as well as the **qualitative** nature of the world. Scientists attempt to answer what, when, why, where, and how questions; they try to understand and describe the universe.

To seek answers to questions, the **scientific method** blends creative thinking, critical thinking, and problem-solving techniques (figure 1.1). First, scientists make **observations.** These then lead to the statement of a **problem** or the asking of a **question.** Scientists formulate **hypothetical** answers to the questions that have been asked about their observations. **Experiments** are then designed and conducted to test the hypothetical answers. The experimenters establish **procedures** to gather quantitative and qualitative **data.** This data is **analyzed,** and a **conclusion** is drawn that supports, modifies, or refutes the hypothetical answer. After repeated experiments lead to the same conclusions, **theories** are formulated. These theories can then be used to explain past observations and to predict answers to new questions.

Science and its quantification of the "stuff" of the universe started about 500 years ago. In the past 100 years, an explosion of scientific inquiry has taken place. Most recently, computers—developed from discoveries made by scientists—are being used to investigate the details of the universe even faster and further away. The inquiry is taking us from the infinitely detailed code of life's all-important hereditary molecule (DNA) to the vastness of space.

Scientific knowledge is neutral; it is neither good nor bad. However, how we use that knowledge can have desirable or undesirable results. Bacteria and viruses can be grown to produce antibiotics or vaccines. They also can be grown to create biological weapons. Chemists create new and useful molecules, but some molecules turn out to be toxic. Physicists explore the mechanisms of matter and motion, and engineers put this information to use for both farm and war machinery.

Technicians use scientific information to make all manner of tools, machines, and buildings—the things we call modern **technology.** Technology is not only an important part of our day-to-day life, it also helps scientists extend their knowledge and understanding (figure 1.2). Thus, we depend on both science and technology to maintain our modern lifestyle.

Look in the index of your science textbook to find the pages discussing the scientific method and the other terms in boldface mentioned above. Compare the more lengthy discussion in your textbook with the basic introduction given here. If you do this, you will have a better understanding of the method of inquiry used to solve problems in our world. Especially important to the process

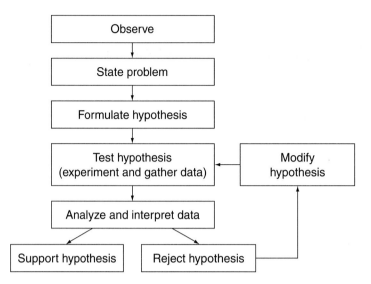

Figure 1.1 The scientific method involves steps of hypothesis formation, experimentation, and analysis. The final step, formulating a theory, is not illustrated in this flow diagram. Theories are developed after repeated experiments lead to the same conclusions.

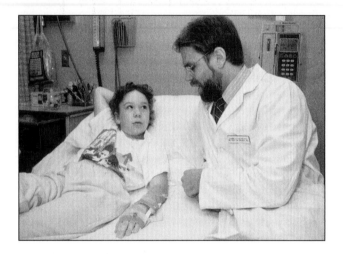

Figure 1.2 The sciences of molecular genetics and medical physiology blend with the technology of gene transfer to enable doctors to give gene therapy. (Courtesy of Dr. Ken Culver, photo by John Crawford, National Institutes of Health.)

are the clear identification of a problem and the formulation of a hypothesis, a proposed explanation to the solution of the problem.

Why Study Science and the Scientific Method?

Since we humans are so dependent on science and technology, we should have some basic knowledge of the concepts and content of science. In addition, we should be able to use scientific information and the scientific method to make decisions about how to conduct our lives.

The welfare of any modern, technological country depends on an informed and active citizenry. Knowledge and communication of science are vital to national issues such as abortion, fetal tissue research, and the future use of genetic code information. Energy use, the production of pollutants, and our desire for greater material wealth are set to converge and create global environmental problems. How can we conduct our lives to resolve or manage these problems? What new legislation should be supported or opposed?

The study of science and the scientific method will also lead to a more effective ability to solve problems in your daily life. For instance, why a car does not start or a computer does not function are problems that need to be solved. Other examples of daily problem solving are finding a job after graduation or managing time and money. All can be resolved by an organized method of analysis.

The following example shows how the scientific method can be used to help manage money. Refer to figure 1.1 to compare this example to a generalized description of the scientific method. As you read through the example, note that details may be lacking, and think of ways that the inquiry could be improved and made more thorough.

Observation: Certain bills are piling up, and debt seems to be larger than income.

Problem: How can you organize payments and minimize interest charges?

Hypothesis: If you change your money management to prioritize the minimization of interest payments and the maintenance of utilities and other basic necessities, you will have income equaling or exceeding expenses.

Test hypothesis:

1. Pay rent, insurance, credit cards on time.
2. Evaluate total income versus expenses.

3. List and reduce nonessential expenses.
4. Determine which bills can be paid late without interest charges or penalties.
5. Don't use credit cards.
6. After three months, recalculate debt and income.

Types of data to be collected: These might include information on expenses, interest rates, and income schedules.

Data analysis: Elements of the analysis might include recalculation of debt and income throughout the three-month period, use of graphs and tables showing expenses, and evaluation of on-going debts and income.

Conclusion: After three months, the recalculation of debt and income will show that income either equals, exceeds, or is lower than expenses. If the income is still lower than expenses, the hypothesis is not supported by your data, and thus either the hypothesis or the procedures must change. If the income equals or exceeds expenses, then the hypothesis has been supported.

Your Study of Science

When we are young, we ask all sorts of questions: What? When? Why? Where? How? Later on, many people stop asking questions; they seemingly stop "wanting to know." But to learn science, you must be ready to ask questions and seek answers.

The learning process applies to all aspects of life, not just school. For example, as a hobby, many children memorize fantastic amounts of data about players from baseball cards. As they mature, some of those youngsters go on to study the strategy of the game. Those individuals have progressed from memorizing to thinking abstractly about concepts—and they have enjoyed doing it! The hobbyist's learning is *self-directed* and *self-disciplined.* Motivation comes from within the person. If this effort could be bottled and sold, it would make millions.

To be successful in your study of science, you must be able to do the following:

1. Identify and locate information to be learned.
2. Organize the information so it can be learned efficiently and effectively.
3. Interpret the spoken, written, and symbolic language of science.
4. Use and apply the information you have learned.

When you study science, you will **memorize** concrete bits of information and learn to **comprehend** complex relationships. This combination of memorizing and comprehending will lead to **understanding.** If you can moti-

vate yourself to learn the material in your science course, you will be successful.

Some students have the skills necessary to do well. Other students have not developed, or have forgotten, the study skills they need to succeed. Still other students lack the self-confidence to study science; they have "science anxiety."

Overcoming Science Anxiety

Some students become so anxious about studying science or taking a test that "their minds go blank" or "they can't think straight." Others feel the course required by the curriculum will make or break their college careers. If you are extremely anxious about a science course, there are at least three things you can do.

The first is to visit a counselor on campus to help you deal with your real but inhibiting anxiety. Many colleges have counselors who specialize in anxiety and stress management. Visiting them could help.

The second thing to do is to concentrate on *learning the skills* and *taking the time* necessary to study science. Do not dwell on the reasons why you can't study, can't learn, or can't take tests. This negative "whirlpool" can consume excessive time and energy. Practice and use the study methods recommended in this book. Begin to prepare for tests on the first day of class. Realize that if you do this conscientiously, you should feel confident about your ability to pass the course. If you learn from the first test, as suggested in chapter 15, you should do better with each passing week and each succeeding test.

A third way to help overcome science anxiety is to seek out or form a **supportive study group.** Find other people who want to succeed in their courses. Studying together does work because it gives you the opportunity to teach each other. It also can be fun.

From time to time, you might need a "crutch" to help reduce your level of anxiety. In 1966, Joseph Wolpe devised the Subjective Unit of Disturbance Scale (SUDS) to help people gauge and reduce their anxiety. Here's how it works:

Using the Subjective Unit of Disturbance Scale (SUDS) Think of two experiences or states of mind that yield two extremely different levels of anxiety. The first is an anxiety-free state, a scene that yields a peaceful frame of mind. Sitting by a shaded, quiet brook or watching a butterfly peacefully feeding on nectar are examples of this mindset. Register this as "0" on the SUDS. Next, think of the most anxious experience or scene that you can imagine . . . for example, a dark alley with a charging, snarling mad dog racing at you. This "worst of worst" experience should be "100" on the SUDS. From this point on, you can evaluate your anxiety level on your SUDS during class, while studying for tests, or just before tests. Are you closer to 0 or 100? Once you do this, you can take measures to reduce your level of anxiety. Think of the brook or the butterfly. Imagine the flowing water or the flittering flight of the butterfly.

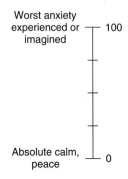

Worst anxiety experienced or imagined — 100

Absolute calm, peace — 0

Learn to think less-anxious thoughts, and use the SUDS to gauge your anxiety. If your anxiety is too high, then take steps to reduce it. Motivate yourself to act positively. This might help reduce your science anxiety enough to give you the confidence you need.

Risk There is a risk involved in studying science or any other subject. A positive self-image and feeling of self-worth will make you feel confident that the risks are really relatively small and can be managed. This is true only if you learn how to learn, study, and be an active learner. The sense of risk can also be reduced if you have the support of peers, spouses, and family.

Meeting the Prerequisites

The college catalog lists the types of courses you must have as prerequisites before studying a science course, but more fundamental prerequisites must also be considered. These fundamental prerequisites combine to create your personal "Climate for Learning" and can either make or break your success in science. It is important to be aware of your climate for learning and modify it to increase your success and satisfaction in science. Figure 1.3 shows the fundamental prerequisites that create your learning climate. You need to have a clear, **personal goal** and the motivation to become **actively involved** in learning (figure 1.3). You also must be willing to make the *effort* to study **6 to 12 hours a week.** How much time you spend studying depends upon your goals, commitment, job requirements, social life, and health. Add to these the requirements of being able to read at the eleventh grade level, having good study skills, and possessing good writing skills. With these prerequisites, plus an understanding of elementary algebra, you will be prepared to take an introductory science course. Students studying in allied health or science major courses need intermediate algebra, college algebra, or calculus.

As you begin a science course, you should evaluate yourself to see if you have the foundation to succeed in the course. If your analysis indicates that you have the prerequisites, you can feel confident about taking the course. On the other hand, if your analysis indicates various deficiencies, you must either strengthen your prerequisites before taking the course or cope with the difficulties and frustrations you will experience. A college

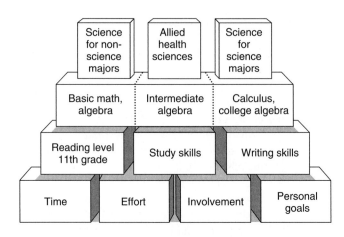

Figure 1.3 Seven "blocks" and an understanding of mathematics are the prerequisites needed to learn science.

Figure 1.4 "You can lead a horse to water but you can't make him drink it." This saying and cartoon relate to you, the subject, and your instructor.

degree is worth achieving. The study of science is one of the basic requirements for such a degree. You can study science successfully and earn your degree if you have proper prerequisites.

If you think you need help in learning the content of a science course, this book is designed for you. Also, if you need help developing study skills, this book will help you. The recommendations are intended to be practical and valuable. They have been gleaned from heeding personal experience, observing and interacting with successful and unsuccessful college students, and reviewing other study skill guides.

Successful students work hard, concentrate on the task at hand, and have an organized system of study skills and a strong foundation of basic skills in reading, writing, and mathematics. With regular practice and application of these study skills based on a strong foundation, you should succeed too!

College instructors assume you are in college to learn. They will lead you to the information, but they don't feel it is their task to get you to learn (figure 1.4).

The Rewards of Studying

A high grade, a high class standing, fulfillment of requirements, a scholarship, recognition, satisfaction of a natural curiosity, and a need to feel competent blend together in a complex way to be your reward for completing a science course. If you are highly motivated, then studying and learning will occur, and rewards will be gained. If you are poorly motivated, the amount of studying and learning will be scanty, and rewards will not materialize. In fact, frustration will replace satisfaction.

You probably do not view the study of a college science course as a hobby. After all, college science courses are requirements. Because courses are requirements, some of the joys a hobbyist experiences while learning might be absent. Nevertheless, try to cultivate an "I can" attitude and an interest in learning. Introductory science courses are challenging, and it is gratifying to do well in them. The study skills recommended in this book will help you.

Review

1. The study of science attempts to define or describe various parts of the universe.
2. The scientific method is an organized system of study to answer questions or solve problems.
3. Humans apply scientific information to develop modern technology.
4. Citizens of technological countries should have an understanding of science and the scientific method.
5. Students of science must be able to ask questions and define problems. They then must be able to devise ways to answer the questions or solve the problems.
6. A person's attitude and academic background influence the way in which he or she will learn and how well he or she will succeed.
7. Varied study skills are needed to learn science.
8. Learning science can be a rewarding experience.

EXERCISE 1

The Study of Science

....

If you need additional space to answer questions, then use a separate sheet of paper.

1. a. Climate for Learning—Self-Evaluation
Your ability to succeed in college science courses is related to the climate in which you are trying to learn. Evaluate your "Climate for Learning" by rating yourself in response to the following statements. Use a scale of 1 (low) to 10 (high) for each statement. Record your evaluation in the spaces provided.

1. b. Reevaluate yourself after three weeks. How has your experience in the science course influenced your evaluation?

Self-evaluation

Week 1 Week 3

_____ _____ 1. I have a positive attitude toward learning in general.

_____ _____ 2. I have a positive attitude toward learning science.

_____ _____ 3. I have a home or dorm environment that makes studying comfortable and possible.

_____ _____ 4. I have the support and encouragement of my family and peers.

_____ _____ 5. I have the motivation necessary to learn science.

_____ _____ 6. I have 6–12 hours per week available to study science.

_____ _____ 7. I can manage my time to complete the things I must do.

_____ _____ 8. I have good note-taking skills.

_____ _____ 9. I have the skill to read and comprehend scientific information.

_____ _____ 10. I have good observational and problem-solving skills.

_____ _____ 11. I have the competency to read, write, and do mathematics at the college level.

_____ _____ 12. I feel good about myself and have confidence in my abilities.

_____ _____ 13. I have the ability to concentrate on a topic for long periods of time.

_____ _____ 14. I have the self-discipline to stick to the task at hand and to not be side-tracked by distractions.

1. c. Improving Your Climate for Learning
If you rated yourself from 1 to 6 on any of the statements in section 1a, those areas need attention. Select three conditions you think you can change, and record two actions you can take to improve your climate for learning.

1. Condition:

_____

_____

Action:

a. _____

b. _____

2. Condition:

_____

_____

Action:

a. _____

b. _____

3. Condition:

_____

_____

Action:

a. _____

b. _____

2. What do you think of the statement, "Scientific knowledge is neutral"?

_____

_____

_____

_____

_____

3. Define and give examples of qualitative and quantitative data. (Use your dictionary if you have trouble defining these terms.)

_____

_____

_____

_____

_____

4. Create a hypothesis explaining why the girl in figure 1.2 has a tube and bandage on her arm.

_____

_____

_____

_____

_____

5. List the requirements to learn science. Refer to figure 1.3.

_____

_____

_____

_____

_____

6. a. Using one of the computer resources in your science course (for example: your textbook's website, CD-ROM, course website), find a definition of the scientific method.

_____

_____

_____

_____

_____

6. b. Explain how the discussion of the scientific method in your textbook compares to the brief statement in this chapter. (See table 13.1, "Characteristics of Essay Tests"). Include the following terms in your essay: observation, problem, hypothesis, experiment, procedure, data, analysis, conclusion, and theory. Use a separate piece of paper for your essay.

6. c. Use your textbook's website to find chapter-related websites for the scientific method. List resources that you found:

_____

_____

_____

_____

7. a. Think of an experience for you that would create a level of anxiety of 0, 25, 50, 75, and 100 on the SUDS.

_____

_____

_____

_____

7. b. Is this SUDS a quantitative or qualitative scale?

_____

_____

_____

_____

8. a. The following observation is made: Students who study in groups seem to get higher grades than those who study alone. Develop a statement of a problem and hypothesis for this observation relating to study method and grades.

_____

_____

_____

_____

8. b. Describe an observation based on your everyday life. Apply the scientific method to this observation and re-state as a problem and hypothesis. Outline how you would apply the scientific method to test this hypothesis. Use a separate piece of paper for your answer.

CHAPTER 2

Science Classes and Instructors

Objectives

When you have read this chapter, you will be able to answer these questions:

1. What are the components of most science courses?
2. How might lecture information be organized?
3. What are the different teaching styles of instructors, and how can I adapt my learning style to each?
4. What is a course syllabus, and how should I use it?
5. Other than class time, is there anything else to help me learn the information?
6. Should I ask questions, and does class participation count?

Your life experience might not give you a clue as to what a science course is about. Thus, you may think taking a science course is riskier than taking other college courses. Part of the challenge is learning the organization of a science course and understanding the style and expectations of the instructors.

How Are Science Courses Organized?

In most colleges, science is taught in a traditional way. Introductory science courses are composed of lecture, laboratory, and in some cases, recitation classes (figure 2.1). Two or three lectures will meet for a total of two-and-a-half to three hours each week. A laboratory period is either a two- or three-hour block of time. A recitation meets for about an hour. The lecture class size may vary from 20 to several hundred students. Laboratory classes also vary in size but generally have one instructor for 15 to 25 students. Variations include lecture and no laboratory or computer tutorials in addition to regular laboratory class.

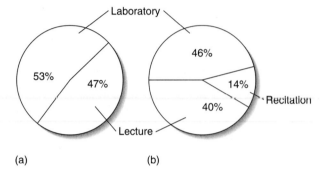

Figure 2.1 An example of a science course. (*a*) This pie graph indicates that lecture occupies 47% and laboratory 53% of total course time. (*b*) If the course includes the three types of classes, recitation occupies 14%, lecture 40%, and laboratory 46% of course time.

Lecture

The bulk of the course content is presented during the lecture by instructors. Instructors use a variety of presentation methods including overhead transparencies, blackboards, photographic slides, and computer projections. The common theme to all these visual aid methods is that key ideas, concepts, and brief outlines are written in the presentation. The main body of the lecture discusses the relationships between these recorded points. *Notes* must be taken on the *discussion* as well as on what is recorded on the *board*. Instructors may also use a variety of other visual aids such as videodiscs, and computer-simulated exercises to present figures and concepts from the textbook. As figures are projected, be sure to record the figure number and make notes on the information described in the visual aid. If lecture notes are provided by the instructor or posted on your course's website, it is still important that you actively read, follow, and supplement these notes before, during, and after the lecture. Note-taking can be an active way to learn information and is a challenging skill to learn. Hints for note-taking are discussed in chapter 5 (Listening and Taking Notes).

Lectures, textbooks, and computer software are organized in a number of different ways. It will help you to listen, read, take notes, and learn if you recognize these patterns. These resources will use different organizational approaches at different times. The following patterns have been organized in outline form. (Creating an **outline** is an important study skill. Note the numbering and indentations of the outline.)

Types of Organization[1]
A. Chronological (in time)
   1. Sequence in which subject was seen or discussed.
      Example: Study of the atom.
   2. Sequence in which the subject was (is) accomplished.
      Example: A laboratory experiment or the solution to a problem.
   3. From cause to effect.
      Example: Radiation causing mutations or cancer.
B. Spatial
   1. What is next to what.
      Example: Different strata of rock in the Grand Canyon.
   2. What is connected to what.
      Example: Description of the digestive system.
C. From general to specific.
   1. Theoretical to practical.
      Example: $F = md$, Work of pulley systems.
   2. General topic to examples.
      Example: Function and nature of enzymes or catalysts to discussion of pepsin or platinum.
D. From least to most (or most to least).
   1. Small to large.
      Example: Atom to biosphere.
   2. Weak to strong.
      Example: Chemical bonding.
   3. Simple to complex.
      Example: Discussion of tissues or the structure of atoms.
   4. Least controversial to more controversial.
      Example: Discussion of origin of the universe.

Laboratory

Laboratories provide hands-on experience with the process of science. The information in the lab might or might not parallel that in the lecture, and the laboratory instructors might or might not be the same as the lecturer. Some laboratory instructors might just sit at the desk and answer questions, assuming that you are doing the work and understanding the material. Other instructors move around

[1] Adapted from *Study Smarts* by Judi Kesselman-Turkel and Franklynn Peterson (Contemporary Books, Inc. 1981).

the lab room to check on your progress and participation. It is up to you to take the initiative to become involved in the lab work. Don't be afraid to call upon instructors for help. It is their job to answer your questions. However, it is a bit embarrassing to ask questions if you have not properly prepared for the class (see chapter 7, "Preview for Laboratory").

You will use laboratory manuals to guide your lab study. Don't sit back and watch your lab partners do all the work. If you do, you will find they will learn the material more easily than you will. Sometimes students with different styles of study and work have trouble working in the same laboratory group. This in turn could interfere with learning. If this problem arises for you, try to resolve it before it interferes with your learning.

It is **important to take notes** on what is covered in the **laboratory class.** If the instructor adds or clarifies information before you begin work on the lab, take notes on what is said!

Laboratory Books and Reports   Some science courses might require you to record procedures and data in lab notebooks and to write laboratory reports (see chapter 12). Careful and organized record-keeping is part of the laboratory exercise. These requirements are reviewed during the first laboratory period. If the requirements are not clear, ask questions. Clarifying what is expected is not stupid; it is smart. The following are things to keep in mind:

*   Know what the instructor expects and requires.
*   Observations and data should be recorded neatly.
*   Graphs, tables, and diagrams should be clearly labeled with captions and units.
*   All calculations must be shown.
*   Incorrect data or calculations should be crossed out with one line. Don't tear out pages or blot out work.
*   Written work (analysis and conclusions) should be concise.
*   Hand in work on time; don't let work pile up.

Recitation/Tutorials

A recitation class is devoted to problem-solving or clarifying information from lecture and lab. Instructors will give assignments one week and review them the next. These activities may include time with computers in a computer lab or tutorial. Students might be asked to demonstrate how they solved the assigned problems. Some instructors give quizzes on lecture or laboratory information during recitation periods. These activities reinforce the content of the lecture and laboratory.

Types of Instructors

As a learner, you have a distinct *style of learning* (see chapter 3, Bridging the Learning Pyramid). Your instructor also

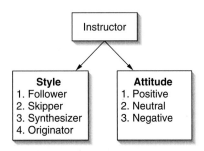

Figure 2.2 Instructors have different styles and attitudes.

has a distinct *teaching style.* Your instructor's style influences the way you listen, take notes, study, and learn. Sometimes, it may seem as though your personal learning style is incompatible with your instructor's style, and vice versa. Recognize that these seeming incompatibilities are not anyone's fault. Equally important to recognize is that there are things you can do to improve your success in the science course by adopting new skills. The first step to adapting to your science course is to identify your own style and the style of your instructor. An instructor's style of teaching can be categorized into one of four groups (figure 2.2): the follower, the skipper, the synthesizer, or the originator. During the semester, your instructor might shift from one style to another.

Here are descriptions of four instructor "styles" and suggestions for ways you as a student can adapt to them:

1. **Follower:** Follows the content of a textbook chapter-by-chapter, page-by-page, example-by-example.
   a. Bring your book to class so you can check the sections of material covered. You should take notes on any additions or clarifications of text material.
   b. You will find it easy to preview material before attending lectures. Your textbook with the checked material and notes are your guides to the material to be studied and learned.
2. **Skipper:** Follows the content of a textbook but skips around from one part of the book to another.
   a. The skipper's content might or might not be easy to trace in the textbook. It's important that you take notes to keep track of exactly what was covered and in what sequence.
   b. You will find it easy to preview material before attending lectures because assigned readings are probably given. For study purposes, your lecture notes are as important as the text.
3. **Synthesizer:** Draws on different parts of the textbook and on outside resources. All is

synthesized into the lecturer's view of the science course. This view might introduce the same material as the text but in a completely different way.
   a. Lecture notes are very important because the topics cannot be quickly found in the text. Study of the notes and of similar information in the text helps solidify learning. Looking up key terms from lectures in the index of the textbook will help you locate information in your text corresponding to the lecture notes.
   b. This type of presentation will be difficult to preview unless the instructor indicates what areas will be covered in the lectures to come.
4. **Originator:** Presents information from a variety of sources, much of it originating from recent publications. Collections of readings might be important, rather than a single text. Independent study and research are expected.
   a. Note-taking is important as is study of assigned outside readings.
   b. If a textbook is recommended, you will probably use it as a reference.

Occasionally an instructor will assume the role of a **facilitator** in the learning process. The instructor helps students define the topics to be learned. Groups of students assume the responsibility of researching and mastering the topics. As this is accomplished, students then take on the role of teacher and teach each other and the entire class. This **collaborative learning** develops responsible academic and work attitudes and shifts educational responsibility to the learner (student). It is a nontraditional, but exciting, way of learning. However, instructors and students "born and raised" to traditional education are uncomfortable with this mode of learning.

All instructors also have a certain *attitude* about their jobs as teachers and about your job as a student. Some instructors display a very positive attitude toward teaching, an enthusiastic interest in the subject matter, and a sincere concern about whether or not you learn the subject. Other instructors are purely dispensers of information and seem to have a very neutral attitude toward their job, the subject matter, and you. Still other instructors are negative individuals; they are dissatisfied with their jobs and would rather be doing something else.

As discussed, you, as a student, have a certain *learning style.* You also have a certain set of attitudes that influence your learning. Your own style of learning might mesh well with your instructor's style of teaching, but if it doesn't, you will have to take steps to cope with the situation. It is your job to learn the material; the college educational system leaves it up to you to comprehend the subject matter. Some instructors make the material seem relatively easy and interesting; other instructors are sources of great frustration.

Course Syllabus and Requirements

Instructors hand out a course syllabus or announce that the syllabus and other materials such as course lecture notes can be found on the Internet. As the instructor reviews the syllabus, have a pencil or pen in hand, follow the review, and take notes. Re-read the syllabus as your first homework assignment. Compare the course outline in the syllabus to the table of contents in your textbook. Recite the topics you will study and reaffirm the requirements for the course. Place the syllabus in your notebook.

The syllabus generally includes the following:

1. Course objectives
2. Title of required textbook and manual
3. Statement of teaching approach
4. Course outline
5. Reading assignments
6. Grading method and values of tests, assignments, and other work
7. Attendance policy
8. Makeup procedures
9. Tutoring center information
10. Computer-assisted learning and website information
11. Policy on course withdrawal

The syllabus is an important part of any class since it tells you course expectations and requirements. Your success in the course relies on knowing and understanding the course syllabus.

Questions and Participation

If you are an active learner, questions will come to mind as you listen in class or as you study. Don't stifle the question. Raise your hand and ask the question or write it down in the question column on the left-hand page of your notebook (see chapter 5). A questioning mind will help define what must be learned. If, for some reason, questions are not encouraged in lecture, be sure to record them in your notes and seek out the answers after class.

Participation in class may or may not be a factor in determining your grade. Just because you attend class does not mean you will pass the course. The course syllabus will give information about the role of class participation and attendance.

Grades

Scientific information will be presented in the classroom in a number of different ways. You are expected to learn the ma-

terial and then demonstrate your comprehension of it. Your responses to tests and written reports result in a grade. You will develop certain feelings about the overall course of study. In the end, you will judge your overall level of satisfaction.

Final grades will be based on an average of test scores and graded assignments. The amount of effort and study time, class participation, and extra projects may be, but seldom are, part of the final grade (see chapters 12 and 13).

You will find that instructors expect you to find out what you missed if you were absent. Your absence does not excuse you from having to learn the material. Exchange telephone numbers with a few classmates so you can call them if you must miss class.

Where to Get Help

Instructor's Office Hours

Instructors generally are required to hold office hours. If you have questions about the course content or requirements, make an appointment to see the instructor. Before the appointment, write down the difficulties or questions you have. This will demonstrate that you've made an attempt to analyze your concerns and will lead to an effective and efficient meeting.

Learning Centers

Most colleges have organized tutoring services to assist you. Find out if a learning center is available for the science course you are taking. Tutors are usually graduate or undergraduate students qualified to help guide your study. As with instructors, their level of skill will vary. Generate questions before you seek their help. The questions will be a valuable way to start a tutoring session. In addition, you or your study group might find the center a good place to study.

Books, Laboratory Manuals, Study Guides, and Computer Resources

A textbook is a primary source of information to complement the presentation of the instructor. Your job as a student is to use the information in the textbook to reinforce the content of the lectures. The laboratory manual will describe the specific exercises you will perform. This manual will contain the bulk of the information you will need to learn in the laboratories.

Most publishers offer study guides to help students learn the information in the textbook. Computers are quickly becoming a popular way of offering study assistance to students. Likely, your textbook and course have websites that contain tutorial assistance, online "virtual" laboratories, web-links to other sites, and chapter-by-chapter exercises and interactive activities. You should listen to your instructor's recommendation concerning the use of the study guides and other references or study aids.

Most textbooks, manuals, study guides, and computer resources contain more information than will be taught in the science course. An important study skill is learning how to identify the content you are expected to learn. Advice on the use of computers will be discussed later. Let the lectures and laboratories be your guide.

Review

1. Introductory science courses have lecture and laboratory classes each week. In addition to lecture and lab, some science courses also have a weekly recitation class.

2. Scientific information will be presented in the classroom in a number of different ways.

3. An instructor might be a follower, a skipper, a synthesizer, an originator, or any combination of these. How you take notes and how you study are influenced by the instructor's style.

4. Instructors have attitudes that influence their effectiveness as teachers. It is your job to comprehend course content regardless of the instructor's teaching style and attitude.

5. A course syllabus describes the course requirements and content.

6. The grades in the science course are based on your performance on various tests and graded assignments.

7. Use instructors and tutors to help you answer questions you can't resolve. Instructors and tutors are paid to help you.

EXERCISE 2

Name _____

Date _____

Science Classes and Instructors

1. List your concerns about taking a science course. Compare your list with classmates. Discuss ways to decrease any potential anxieties.

   _____
   _____
   _____
   _____

2. The flow chart below illustrates various components that will decide your course grade and ultimately your satisfaction in your science course. Describe this figure in short essay form (See table 13.1, "Characteristics of Essay Tests). Use additional paper, if necessary, and compare your essay with the essays of other students.

   _____
   _____
   _____
   _____
   _____
   _____
   _____
   _____

3. What kind of graph appears in figure 2.1? Draw a graph to show how much time each course activity occupies in your science course. Be sure to include

any computer tutorial/laboratory time as a separate piece of the pie.

   _____
   _____
   _____
   _____

4. List the four styles of instructors you might encounter. Which of these styles do you think will match your style of learning?

   a. _____ c. _____

   b. _____ d. _____

5. If you think the style of your instructor will clash with your style of learning, what three actions will you take to learn the science material presented?

   a. _____

   b. _____

   c. _____

6. This chapter discussed a number of ways instructors might organize their lectures.

   a. In what format was "Types of Organization" presented in this chapter?

   _____
   _____

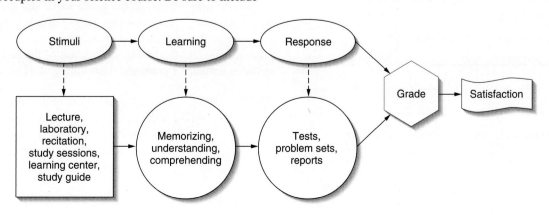

6. b. What are the four main types of lecture organization? Give an example of each type from your own science course.

1. _____   _____
   _____

2. _____   _____
   _____

3. _____   _____
   _____

4. _____   _____
   _____

7. Evaluate the list of things included in a syllabus. List the six that are most important to you.

_____   _____
_____   _____
_____   _____

8. a. Does your course syllabus provide information on how to get extra help? If so, describe how you can get extra help. If not, list actions to take to find out how to get extra help.

_____
_____
_____

8. b. Does your college have a science learning center? If so, where is it on campus, and during what hours is it open? Can groups meet there to study?

_____
_____

9. In the following groups of words, one word does not belong in each group. Circle the word that does not belong, and indicate why you think it does not belong in the group.

a. data, syllabus, hypothesis, analysis

_____
_____
_____

b. experiment, skipper, follower, synthesizer

_____
_____
_____

c. table of contents, procedure, glossary, index

_____
_____
_____

10. a. Explain how your final grade will be determined.

_____

10. b. Place a data point for yourself on the graph that follows.

Graph of student grade vs. satisfaction in a science course.

Note that if you answered question 10, you had to

- Read the labels on each axis.
- Choose a level of satisfaction.
- Choose a grade.
- Locate the level of satisfaction on the vertical scale and locate the grade you desire on the horizontal scale, then draw an imaginary line from the satisfaction axis and up from the grade axis. Where these lines meet is where you placed yourself on the graph.

Generally, the degree of self-satisfaction a person feels is directly proportional to the grade he or she earns. The higher the grade, the greater your feeling of self-satisfaction. However, exceptions do exist. The data points represent real student data, and the line is a prediction of the relationship between a student's grade and the resulting satisfaction level.

10. c. What actions will you take to ensure that you are satisfied with your grade in your science course? Use a separate piece of paper and be specific. Discuss your answer with your classmates.

CHAPTER 3

Bridging the Learning Pyramid

Objectives

When you have read this chapter, you will be able to answer these questions:

1. How does a learning pyramid apply to my learning?
2. Why isn't just going over my notes good enough to earn a good grade?
3. Can't I just memorize everything?
4. What's the difference if I think with my left or with my right brain?
5. What study skills can I use to help me learn science?

The intent of this book is not to examine the complexity of learning but to introduce ideas about how to learn and to suggest skills to help learn the information presented in science.

All too often, students seem overwhelmed by the content of science courses. Part of the problem is that your life experiences might not help you in science. The vocabulary is all new, and the relationships described are complex and often multidimensional. Instructors seem to spend forever on the details of a subject, and it is difficult to keep track of it all. The textbook is so dense with information that it is difficult to pick out what should be learned. In addition, you might not be able to study efficiently and effectively. Sometimes students don't recognize that they don't know something.

All of these frustrations culminate in questions like the following:

- "Will this be on the test?"
- "Am I responsible to learn that?"
- "Can we have a review before the test?"
- "Will you tell us what we have to know?"
- "What can I do for extra credit?"

The Learning Pyramid

When you listen to a lecture, you tend to forget the information you hear very quickly. If you read something straight through, you tend to forget the beginning of the text before getting to the end. In other words, you tend to learn or remember relatively little of what you hear or read. How many times have you left a lecture or finished reading a chapter and said, "What did he or she say?" or "What was in that chapter?"

You learn a bit more effectively if the information is enhanced with either audiovisual aids or demonstrations. If, on the other hand, information is conveyed by a hands-on experience, your learning tends to be even more effective. The combination of seeing, listening, and doing seems to form more lasting memories. These, in turn, can nourish the thought processes that are involved in understanding information.

Computers are tools that combine listening, reading, and graphic representations. In addition, interactive programs stimulate involvement in all of these. Thus the computer could be a valuable tool to help you learn science. However this is only true if you develop learning objectives and use the computer to fulfill the objectives. Be careful not to browse and search software or the Internet learning fascinating information that has no relationship to your course requirements and objectives. For example, don't use computer lab time to "surf the internet" assuming that you can do the work at home. Use your computer tutorial/lab time to work on course exercises.

You tend to learn and understand information best when you teach it to someone. In preparing to teach, you must think about and evaluate the information you are teaching. In addition, the teaching process forces you to practice using and connecting the concepts you are learning into complete thoughts expressed in terminology you understand. Thus, a **learning pyramid** can be constructed from these learning experiences (figure 3.1). The most learning takes place when you use and teach the information. The least is learned by reading or listening (table 3.1).

Most science on college campuses is taught through lectures and readings, but this is the least effective way to learn. The use of transparencies, films, and videos might increase learning, but all too often instructors do not use

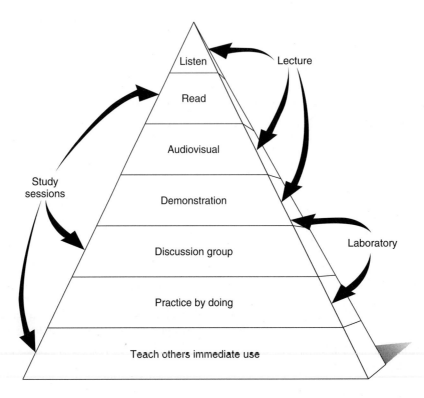

Figure 3.1 The learning pyramid. You learn and retain information best when you use it immediately or teach it to someone else. You learn least effectively by just listening and reading. You must use various study skills to connect the way information is presented and the way you learn most effectively and efficiently.

Table 3.1 How We Best Learn, based on "*The quality school: Managing students without coercion*" by William Glaser (Harper Collins, 1992).

We learn:

- 10% of what we read
- 20% of what we hear
- 30% of what we see
- 50% of what we see and hear
- 70% of what is discussed with others
- 80% of what we experience personally
- 95% of what we teach to someone else

these learning aids effectively. Laboratories set a stage for "learning by doing," but many students do not take the initiative to prepare and participate in the laboratory exercise. The idea of trying to teach the information you are struggling to learn never occurs to some students. A component of teaching is involved in collaborative learning, but this learning mode is rarely found on the college campus.

The answer to effective learning is, in fact, housed in a learning pyramid. You must select and use a variety of study skills that will bridge the top of the learning pyramid with the bottom of the pyramid. In other words, you must

engage in study activities that will allow you to use and teach the information you have heard and read. It is your responsibility to build this connection between the style of education in most colleges and the way in which you can learn efficiently and effectively. Each person's blend of study tools will be a bit different.

In the end, you will have to develop study skills, learn how to learn, spend the time, and make the effort.

Memorizing and Understanding

Some of the material you will be required to learn will involve memorizing specific or concrete bits of information. Some examples are the names of elements, atomic numbers and weights, molecular structures and formulae, cell types and cell parts, and the names of minerals and rocks. Definitions of terms may even need to be memorized. If questions are asked in the same way you have memorized information, your answers will be right if you can recall the things you memorized. If you can't recall correctly, or if things get "mixed up," your answers will be wrong.

Other material that must be learned will depend upon more abstract thought. Although you will need to memorize the terms, it will be expected that you'll be able to arrange those terms and concepts into sequences that make sense. Comprehending something will mean that you will be able to use and synthesize information. As you connect concepts, you'll be expected to place concepts in hierarchical arrangements.

Here are two examples to show the difference between *concrete learning* by memorization and *abstract learning* that involves more thought and understanding. In an anatomy course, you will need to memorize the names and locations of specific muscles. This is rather straightforward: see it, memorize it, and recall the information. The difficulty might be the amount of information you need to memorize. By contrast, in physiology, the chemical nature of muscle contraction would be discussed. Physiological information about molecules and their interaction are more abstract. The sequences of processes and structures that interact are complex. Memorization helps you learn some of the information, but the ability to think abstractly and to make interconnections is required to understand what has to be learned about the control and coordination of muscle action.

If you were asked to relate force, resistance, and effort to a math-based problem on muscle strength, the principles of physics and the concepts of mathematics would have to be applied. This type of problem-solving and mathematical evaluation is more abstract than the memorization of the name and location of a set of muscles.

In laboratories, you will be doing various kinds of exercises and experiments. It is relatively easy to memorize the names of taxonomic groups, chemical compounds, and structures of organisms. However, students find it difficult to compose discussions and conclusions in laboratory reports. These segments require thinking about the experiment and related information. In other words, just knowing individual terms is not enough to demonstrate that you understand the information.

You and Your Learning Style

Your unique learning style is made up of a personal set of conditions under which you learn best. As you become more aware of your learning style, you can incorporate new learning tools into your academic and personal life and be more successful in learning. For example, if you feel that your style is incompatible with learning science, there are steps and actions that you can take to better learn science. Learning styles are described in different ways. One way is "Left Brain/Right Brain."

Left Brain/Right Brain

The right and left hemispheres of our brain process information differently. The left hemisphere "sees" things in linear order and processes information in a sequential manner. The right half tends to process information in a global fashion. Each person blends the actions of the two hemispheres as he or she interprets and interacts with the world. Some people are *integrated* in the way they learn things. That is,

Table 3.2 Brain Dominance Characteristics

Left Hemisphere	Right Hemisphere
Sequential and systematic	Random learning and thinking
Structured and orderly	Relies on images or visualization
Makes list	Lists are not important
Plans out tasks	Tends to let things fall together
Tends to keep "stuff" in order	Intuitive by nature
Follows directions carefully	Prefers to see demonstrations
"Sees the trees in the forest"	"Sees the forest but not the trees"
Pays attention to detail	Doesn't like to read detailed directions
Sees cause and effects	Likes subjective rather than objective tests
Schedule and time are important	Time and schedules not priority
Likes to feel in control	Likes brainstorming
Dislikes improvising	Flexible in work

they use both sides of their brain to process information. Other people might have a tendency to be either "right-brained" or "leftbrained." Left and right hemisphere characteristics are given in table 3.2.

Science is taught in a fashion that favors the left hemisphere or an integrated brain. Information is taught in sequences and hierarchies. Many laboratory procedures follow a specific sequence. Instructors organize their lectures in sequences (chapter 2). If you have a preference for right-hemisphere thinking and action, you may find learning science difficult or frustrating. How can you begin to see and learn sequences? How will you get yourself to identify the details you have to learn? Recognizing the type of brain dominance you have could allow you to take measures to adapt your learning style to become a more integrated thinker. For example, if you find that you are right-hemisphere dominant, you may need to develop list-making habits to better structure your time.

Study Skills

Many different study skills will help you learn science. Remember, the objective is to connect the top of the learning pyramid to the base of the pyramid. You should choose a variety of study skills and learn how to use them. When woodcarvers first learn how to use different tools, they need to think about how they use them. As they become expert carvers, they no longer need to think about how they hold or use their tools. Your use of different study skills, your tools to learn science, should likewise become second nature to you.

Your class activities must be integrated with and re-inforced by study outside of class. To do a good job, this takes time and a variety of study skills. Most college instructors feel that for every hour spent in class two hours should be spent in study sessions. You should select and develop a set of skills to put in your "learning toolbox."

Some students just review notes many times and think that is studying. Their experience might be based on prior learning experience. Their teachers "gave" them notes and then tested the students on the notes that were given. This mode of teaching did not encourage independent note-taking or the development of note-taking and listening skills. In addition, memorizing the notes worked because tests could be passed.

Not so in college. In most college courses, notes are not "given"; they must be taken accurately by the student without too much help from the teachers. They must be enhanced and corrected outside of class. Tests probe and measure understanding and thinking rather than your ability to memorize.

Grades you earn are a measure of how well you learn or how well your study skills work. Poor grades mean you have not met the standards of the course. Spending more time studying in the same way probably will not improve your learning. Reviewing notes over and over again is not enough since your notes might be incomplete and inaccurate. You have to consider adding other study skills to help you learn more effectively.

The study skills you put into your "learning toolbox" relate to your general attitude and behavior, to your activities in class, and to your study outside of class. These skills are discussed in various parts of this book, and the end-of-chapter exercises help you develop the skills. Remember, you must start to practice a variety of study skills to actually learn them. You must really learn how to learn! Important ingredients of your attitude and behavior, in-class skills, and study outside of class are as follows:

General Attitude and Behavior

1. Commit yourself to a specific study schedule.
2. Develop an "I will" and "I can" attitude.
3. Be an active learner.
4. Develop a sense of self-direction and self-discipline.
5. Incorporate science into your life.

In-Class Skills

1. Attend all classes.
2. Listen with care and attention.
3. Take complete and accurate notes in lecture and lab.
4. Use a right-hand page, left-hand page format.
5. Record questions asked by the instructor.
6. Participate actively in laboratory and recitation classes.
7. Develop good test-taking skills.

Study Outside of Class

1. Create "to do" lists.
2. Establish learning and reading objectives.
3. Study 6 to 12 hours each week.
4. Preview and review for all classes.
5. Comprehend what you read.
6. Prepare summaries.
7. Generate questions for self-testing.
8. Develop a system to learn terms.
9. Use a study guide.
10. Seek help from the tutor or instructor.
11. Complete all assignments.
12. Analyze cause of error on tests and reports.
13. Study in a group to teach each other.

Review

1. A person tends to learn relatively little from listening to lectures and reading books.
2. A person tends to learn more using and teaching the information that is being learned.
3. Some things must be memorized.
4. Abstract thought is needed to relate, evaluate, and synthesize information. This will allow people to demonstrate they understand the information.
5. People might process information by using the left or right hemispheres of their brain.
6. In most colleges, science is taught in a "left-brain" fashion.
7. A variety of study skills can be used during classes or outside of classes.

EXERCISE 3

Bridging the Learning Pyramid

....

1. a. Check which of the following behaviors apply to you.

_____ I tend to look at individual flowers rather than the bouquet.

_____ I read the detailed directions before I attempt to put anything together.

_____ I tape my class schedule on the inside cover of my notebook.

_____ My money is scattered about: some in my pocket, purse, or wallet but not in order.

_____ My $1, $5, and $10 dollar bills are in order in my wallet.

_____ My books are in my room, somewhere.

_____ My books are all in order, and I'm angry if someone disturbs that order.

_____ I'm always on time.

_____ I'm hardly ever on time.

_____ My room looks like it took a direct hit by a bomb.

_____ My room is orderly with everything in its place.

1. b. After comparing this list to table 3.2, decide whether you are:

_____ right-brain dominant,

_____ left-brain dominant, or

_____ integrated-brain dominant (both right and left hemispheres)

1. c. Refer back to chapter 2. Now that you know more about your own learning style, do you still agree with your assessment in questions 4 and 5 from chapter 2? If no, why not?

_____

_____

Now that you have started to describe your learning style, be sure to complete exercise 7.

2. a. Evaluate the idea of the learning pyramid and relate it to your experience of studying and learning. Do you think the pyramid applies to you? If not, why not?

_____

_____

_____

_____

_____

_____

2. b. Explain how you might use computer resources in the learning pyramid and with what levels of the pyramid the different computer resources connect (for example, computer tutorials/laboratories, web-based study guides, online tutorials).

_____

_____

_____

3. List four specific study skills or activities that you think will bridge the base to the top of the learning pyramid.

_____

_____

_____

_____

4. What do you do in between tests to monitor your progress in learning the course content?

_____

_____

_____

_____

_____

5. Studying and thinking are obviously related. Think about the words representing the concepts below. Match the words in column 2 to the words in column 1.

Column 1	Column 2
_____ Memorize	a. Create
_____ Understand	b. Evaluate
_____ Use	c. Analyze
_____ Thinking in sequences	d. Apply
_____ Determine right, wrong, fact, fiction	e. Comprehend
_____ Synthesize	f. Know

6. List the levels of the learning pyramid. Give a specific example of an activity for each level.

_____

_____

_____

_____

_____

7. a. An inventory of study skills will help you identify the skills you want to have in your "learning toolbox." First place an "x" in column 1 next to the "skills" that you already use. Then place a " + " in column 2 next to the new skills you think would help you study and learn more effectively and efficiently.

7. b. After a few weeks, review this list. Once again indicate what skills you are now using. What new skills are you finding helpful and have become part of your study habits?

Study Skills Inventory

1	2	Study Skills or Activities
_____	_____	Construct a semester calendar of course requirements.
_____	_____	Commit yourself to a specific study schedule.
_____	_____	Study science 6 to 12 hours per week, every week.
_____	_____	Make a "to do" objective list for what is to be studied.
_____	_____	Develop an "I will" and "I can" attitude.
_____	_____	Preview information to be covered in lecture and laboratory.
_____	_____	analyze text and laboratory manual using S + Q + 3R + P (see chapter 8, Use of Textbooks).
_____	_____	Create your own word list from a survey of the textbook.
_____	_____	Get to class on time; quickly review the previous lecture.
_____	_____	Listen with care and attention.
_____	_____	Take complete and accurate notes in lecture and laboratory.
_____	_____	Review, correct, and add to notes taken in class.
_____	_____	Summarize each lecture with an essay, concept map, or outline.
_____	_____	Generate questions for self-testing.
_____	_____	Recite, recall, and envision the material studied.
_____	_____	Use figures to help understand and comprehend information.
_____	_____	Construct scientific figures explaining what you have read.
_____	_____	Keep up with all assignments; don't cram or procrastinate.
_____	_____	Practice solving questions and writing brief essays.
_____	_____	Use the correct format for lab reports.
_____	_____	Construct a bank of flash cards.
_____	_____	Test yourself with questions from the textbook or manual.
_____	_____	Use a study guide.
_____	_____	Study with a study group.
_____	_____	Seek help from a tutor or instructor.
_____	_____	Develop good test-taking skills.
_____	_____	Analyze mistakes on tests and graded assignments.
_____	_____	Relate the science you study to the things you read and see.
_____	_____	Use available computer software programs.
_____	_____	Tell other people what you have studied.
_____	_____	Spend computer lab/tutorial time on work, not play.

CHAPTER 4

The First Week and the "Two D's"

Objectives

When you have read this chapter, you will be able to answer these questions:

1. How can I get off to a good start on the first day of class?
2. Why should I get to know my classmates?
3. What does my instructor expect of me?
4. What are the "two D's," and how can they help me succeed in my science course?

All sorts of questions can flood a student's mind as he or she enters class on the first day. Will this course be rough? What will the instructors be like? What will be required? Will other students be better prepared? Can I ever understand the information? Will I pass? The answers to these and other questions will take the whole semester to answer. However, the first day is important because it can set the tone for the entire term.

Front and Center

You must be an involved active learner in lecture. One way to force involvement is to try to sit front and center. Of course, this isn't always possible. In lecture, always be certain to do the following:

1. concentrate on listening;
2. be in a place where you can see the boards and screen;
3. don't be distracted by fellow classmates;
4. don't fall asleep;
5. receive handouts and tests as soon as possible.

Start learning the course material at the beginning of the term. Don't leave it until the first test or the end of term. For the first and all subsequent lecture classes, you should bring pencils and pens, a notebook (note paper), and your textbook. If you can get to class a few minutes early, open your notebook and get ready to start before your instructor begins. As the semester continues, you will be able to use these few moments to review your notes or even compare your notes with those of other students. For lab, bring pen, pencil, notebook, textbook, and lab manual. As the term progresses, you might find that you do not need the text, but it is always a handy reference to have on hand.

If, on the first day, your instructor presents an overview of the course, *take notes.* Don't just sit back and listen, assuming that this summary is unimportant or that you will remember it. You can gain some important clues about the entire course on the first day. In addition, your instructor will start to display his or her style and standards. You can begin to practice your listening and note-taking skills. You can begin to gauge what kind of challenge the science course will be. After the first class, review the syllabus. Compare the course topics to the table of contents in the book. Get some idea of how much of the book will be studied. Flip through the chapters to be studied to become familiar with their organization and content.

Meeting Your Classmates

Start to get acquainted with your **classmates.** Open a conversation with them. Determine who might have **similar goals** and **attitudes** toward study. Begin to find a group of people interested in forming a study group. This group work is refreshing and rewarding. If you are outgoing, this will be easy to do. However, if you're somewhat shy, it will take motivation and confidence on your part to come out of your shell. Remember, group study is another important study skill.

Some students are intimidated by the behavior or apparent knowledge of other students. Try not to be intimidated. Do your own thing. Study and learn. Take control of yourself and your actions.

Understanding Your Instructor's Expectations

Instructors expect self-disciplined and self-directed study. Reading assignments are given. It is up to you to decide when and how thoroughly you study the assigned and unassigned material. Often instructors will say, "Answer the questions at the end of the chapter or at the end of the lab exercise." That is the last you might hear about the assignment. You are just expected to take the initiative to answer the questions. The next time you see the questions, they might be on a test.

Instructors assume you will do the following:

1. know how to study;
2. have an appropriate academic background;
3. study until you learn the course content;
4. know how to listen;
5. seek help if you need it;
6. know how to determine what to study;
7. be self-motivated;
8. be self-disciplined; and
9. be self-directed.

The "Two D's": Discipline and Direction

Students indicate that two difficult adjustments to college life involve self-discipline and self-direction. Academic **self-discipline** is the ability to establish and follow a set of rules and regulations to guide one's academic behavior. Most students have no problem behaving in a socially acceptable fashion. The big problem is how to behave in an academically acceptable fashion. As a student, you must develop study rules and stick to them. This requires self-discipline. No parents or teachers will remind you. You have to motivate yourself to do the work. You have to set the alarm clock and get up. You have to go to class and ac-tively listen and learn. You have to arrange to make up missed work. You have to set the time to study efficiently and effectively.

Self-direction is the ability to decide when, what, where, and how much you should do. It is a behavior pattern established by yourself for yourself. It is a pattern that directs you to establish certain objectives and allows you to accomplish and achieve the objectives. If you have trouble directing yourself, find out how others direct themselves. Ask how they give themselves direction to study. If you join or form a study group, the group will tend to give you direction. Don't be afraid to seek help and encouragement from your peers.

Review

1. Being attentive in class is the first step to succeeding. Just attending is not sufficient.
2. An involved learner will be prepared with pens, notebooks, textbooks, and lab manuals.
3. Listening carefully and taking notes on the first day of class will give you insight into the course content and help you begin to know the instructor's style.
4. By getting to know other students and forming a study group, you will be using one more study skill.
5. It is your responsibility to learn the material presented in class and to complete assignments. You should seek help from other students or the instructor if you need it.
6. You must establish academic self-discipline rules to guide behavior such as getting to class on time, learning course material, and making up missed assignments.
7. Self-direction is the ability to decide when, where, what, and how much you must do in order to achieve your objectives.

EXERCISE 4

Name _____

Date _____

The First Week and the "Two D's"

....

1. After the first day of class, complete the following Science Course Survey. If you can't decide on an answer, review the course syllabus or discuss that point with a classmate or the instructor. After two weeks of classes, redo the survey.

Science Course Survey

a. How many lecture and lab hours do you have each week?

_____

b. How many hours do you predict you should study each week?

_____

c. How will your grade be determined?

_____

_____

d. What is your instructor's teaching style? (see chapter 2, "Science Classes and Instructors")

_____

e. Is there a limit to the number of absences you may have in lecture? In lab? What happens if you exceed this number?

_____

_____

f. What kinds of laboratory assignments are given?

_____

_____

g. If a laboratory report is required, has the format of the report been described? _____ Is the format clear to you? _____ If not, how will you determine the acceptable format?

_____

_____

h. Where and when are tutoring services available?

_____

i. What are your instructor's office hours?

_____

j. Can you contact your instructor by e-mail? If so, are there any limitations on access?

_____

k. Does the course outline include detailed reading and computer-based assignments? _____ If not, how will you know what to study?

_____

_____

l. What is the website for your textbook?

_____

m. What information can be found on the textbook's website to help you study for the course?

_____

_____

n. If your science course has a website, what is your course's website address?

_____

o. where is your syllabus found on the website?

_____

p. where do you find your grades on the website?

_____

2. Label which of the following statements require self-discipline and which require self-direction.

a. Getting up in the morning for an 8:00 class.

_____

b. Checking the spelling of scientific terms in your class notes.

_____

c. Redrawing a diagram recorded in your class notes.

_____

d. Staying alert and attentive during the last five minutes of lecture.

_____

e. Making diagrams or flow charts of the procedures to be followed in laboratory.

_____

f. Studying for 6 to 12 hours spread out over the period of a week.

_____

g. Reworking the lecture notes either into summary essays, concept maps, or outlines.

_____

h. Just saying "No" to an invitation to go for a beer during a night for study.

_____

i. Meeting regularly with other students to study.

_____

j. Explaining what the lecture notes say during a study group.

_____

k. Meeting with the instructor or tutor to clear up questions you have about lecture material.

_____

3. As you begin the semester, problems not directly related to the study of science may interfere with the study of science. Check the following problems that pose a challenge to you and think how you will meet these challenges.

_____ a. Conflicts of values with peers and friends.

_____ b. Not knowing what to expect and what is expected of you.

_____ c. Adjusting to dormitory life.

_____ d. Adjusting to responsibility for your own actions.

_____ e. Adjusting to an independent lifestyle while still living at home.

_____ f. Making your own decisions.

_____ g. Missing friends and family.

_____ h. Not enough money.

_____ i. An outside job.

_____ j. Time—too much or too little.

_____ k. Moral support of family, spouse, and friends.

_____ l. List other challenges you may face.

_____

_____

_____

_____

CHAPTER 5

Listening and Taking Notes

Objectives

When you have read this chapter, you will be able to answer these questions:

1. How can I become a better listener?
2. Why should I take notes in lecture and lab classes?
3. What techniques can help me take better notes?

An editorial in a journal for science teachers states, "When they come to us, they don't know how to read and they don't know how to listen . . . Similarly with the lecture, you must take working notes that copy at least what's on the chalkboard. Then within 24 hours you must rewrite the lecture, noting points you don't understand."[1] Most science instructors would agree that students need to have good listening skills and must be able to create a notebook that is a learning tool.

Listening

Listening, observing, and writing are three class activities that reinforce each other to allow you to learn. Listening is an important study skill. Here are some tips for listening in class:

1. Don't be judgmental. Your objective is to learn the content and concepts of the presentation.
2. Enjoy a good lecture, but learn to tolerate a poorly delivered lecture. Don't forget to take notes during either type of lecture.
3. Don't be distracted by a lecturer's repeated "Uuuh's," "Okays," and "You know."
4. A brief preview will help prepare you to listen better. Review your notes and preview your text.
5. Be willing to change old concepts and attitudes.

6. Keep your mind open to new ideas and concepts.
7. Realize that, as you listen, college science will increase the depth and breadth of knowledge you already have.
8. Tune out distractions . . . stray thoughts, problems, anxieties. This is easy to say but difficult to do.
9. Extract the broad concepts, then fortify these with specific information.
10. Be attuned to the gestures, tone, and body language of the instructor.
11. Be physically and mentally alert.

Realize that, as you become a better listener, class time will actually be "learning time." It is your responsibility to listen in class. College instructors do not necessarily feel it is their responsibility to get you to listen. Of course, many instructors do use humor, odd facts, and interesting tales to attempt to motivate students to listen.

Taking Notes

Your observational skills in the class are important. Seeing the words written on the board, observing the details of figures, and detecting the body language of the instructor are study skills. Observing these in the classroom will reinforce what you are hearing. Writing notes helps transfer what you have heard and seen to the written word. Seeing the written words provides visual learning.

Notes taken in lecture and laboratory are the primary guide to what you should learn. **Remember, 90 to 100 percent of science tests are based on material covered in class.** If you have not recorded what is covered in class (both in lecture and laboratory), your studying will be inefficient and ineffective. Be sure to record both the material discussed by the instructor and the information written on the board. Even though you think you will remember or recognize what was said in class, experience and experiments indicate that you will not (see figure 7.2).

[1] From *The Physics Teacher,* volume 33, October 1995.

Notes should be a record of what is covered in class, including details and emphases presented by the instructor and your own questions about the material. Many courses now offer lecture notes and presentations on course websites to supplement the lecture. Students can make copies of the notes and presentation before coming to class. Keep in mind that information on the length of time spent and emphasis on topics can only be gained from attending lecture. If your course has online lecture notes, it's important to have these notes. However, you must be prepared to add to them and fashion your lecture notes to reflect the presentation in class in order to make the notes effective and efficient learning tools.

People with poor listening skills or with too much on their minds tend to let their minds wander. Physically you might be in the classroom, but mentally you're on vacation somewhere, dreaming about someone, or worrying about something. If you realize this happens, you must do two things: train yourself to go on fewer mental trips and leave a space in your notebook to indicate you've missed some of the lecture.

The notes you take should be **corrected, clarified, and enriched** during study sessions. Taking notes and crafting them into a study tool is a study skill. It takes time, work, and practice to develop and improve this skill. Note-taking is an active study skill, which increases your ability to memorize and understand. Perceive the development of this skill as a job where there is about two hours of work per one hour of lecture. Remember, you are training yourself for a new way of thinking. If you do improve your note-taking skills, you will increase your chances of successfully learning science. You can judge your note-taking skills by filling in the "Class Notes Checklist" at the end of this chapter.

A Cornell University student started a note-taking business to provide quality notes to thousands of subscribers. Since people are willing to pay for good notes, they must be important.

Choosing a Notebook

What kind of notebook is best for class notes? Spiral-bound notebooks are easy to carry and more convenient to use on classroom desks. However, since you can't add pages to spiral notebooks, you may be forced to keep information relating to the same topic in different places. Loose-leaf notebooks have the advantage that course handouts, study sheets you create, or returned tests can be placed next to similar information in your notes. They might not be convenient in class, but there is no reason to carry the loose-leaf book to class. Just carry the paper to class in a folder and then place the notes in the book as you begin to review them. If you do this, don't forget to number and date the pages.

Formatting Your Notebook— Right Page/Left Page

The major improvement most people can make to their notes is formatting the notebook pages and leaving space. A notebook is the cheapest part of a college education, yet many people are very frugal with the use of paper, cramming stuff in and leaving little room to modify and enrich their notes. If you have problems taking notes, if you feel anxiety about note-taking, or if you have never thought about the importance of notes, read the following pages very carefully! We describe important note-taking skills. If you feel confident about the quality of your skills, then read this section to learn new note-taking habits.

Keep in mind that exams typically focus on lecture material, so good notes are essential. This note-taking style can be adapted to other types of notes such as course notes provided by your instructor or notes written with a laptop computer. If you decide to rely on other methods, be sure to include the important points of note-taking style described in the next sections of this chapter.

It is strongly recommended that you record class notes **only** on the **right-hand page.** Leave the **left-hand page** blank and use that space for additions, corrections, and enrichments to the right-hand page (figure 5.1).

Using the Right-Hand Page

The right-hand page should be used to do the following:

1. Record the date.
2. Record assignments.
3. Review: Some instructors make it a habit to review the important concepts of the previous lecture. If so, then take diligent notes on the review. These points may well be seen on a test at a later date.
4. Summary: This could be placed at the beginning or at the end of the day's notes. If you decide to place it at the beginning, then you must leave blank space for the summary. After reviewing the notes, write a concise summary of the notes. Just remember that writing the summary will force you to reformulate information you have learned in your own words (See figure 5.2).
5. Record your class notes in simple phrases or in outline form. Avoid recording every word the instructor says. How you do this depends on your present note-taking style and your instructor's teaching style. Indentions help separate minor points from major ideas. Dashed lines and numbers or letters also help. Follow these tips:
    a. Be consistent in your style.
    b. Write legibly.
    c. Leave spaces for inserting missed material or clarifying confusing material.
    d. Copy information from the board and overhead transparencies accurately.
    e. Record key terms and page references in the margin.

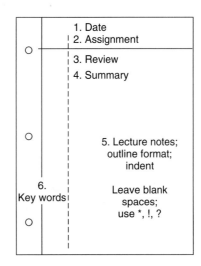

Figure 5.1 Organization of a notebook. Create a tool rather than a "jumble." Use the right-hand page for class notes and the left-hand page for questions, corrections, and enrichments. As discussed in the text, this style can be adapted to other forms of note-taking.

6. Key words from the notes could be listed in the margin during lecture and as the notes are reviewed.

Other hints or ideas that are worth considering to improve your note-taking skills and help you create a learning tool rather than a jumble of recordings include:

1. Use *, !, and ? to identify important points.
2. The list of key terms in the margin might really be called test words. During your reviews, these words should be incorporated into a vocabulary list or flash card system. Remember, the key terms are potential test words.
3. Develop your own shorthand. Eliminate vowels and use symbols. "Develop" becomes "dvlp"; "and" becomes "&"; "turns to" becomes "⟶."
4. Use tape recordings only to check or clarify notes, not to listen to the whole lecture again.
5. Note whether your instructor repeats material in different ways or gives different examples concerning the same topic.
6. Record clues given by the instructor. The tone of voice, the use of expressions, and various gestures are all important parts of body language. Use !! or ** to record nonverbal means of emphasis.
7. Be alert for indicator words or phrases that signal information you should record in your notes. Examples include:
   a. There are three important . . .
   b. The beauty of this is . . .
   c. Therefore . . .
   d. In summary . . .
8. Attend all lectures. Don't depend on other people's notes.
9. Be sure to take notes on the discussion as well as on what is written on the board.

Record the examples given to demonstrate the concepts.
10. If figures are shown on transparencies or slides, make as quick and simple a diagram as possible. If the transparency is a figure in the book, record the figure number and page number. Look at the figure in the textbook as the instructor explains it. Place ! or * next to the things covered by the instructor. Enrich and correct your diagrams during review study sessions.
11. Be sure to copy diagrams from the board accurately or rework a new diagram on the left page.
12. Indicate in your notes as to how much time the instructor spent on a topic.
13. Preview textbook material and the previous day's notes before attending the next class.
14. Review and revise class notes as soon after lecture as possible.
15. Keep notes simple but complete.

Using the Left-Hand Page

Leave the left page of your notebook blank during the lecture, but use it during your review study session. Divide the left page into two columns, one occupying one-third, and the other occupying two-thirds of the page. The two-thirds column could be used for corrections and enrichments. The one-third column could be the question column (see figure 5.2).

Corrections and Enrichment Column The corrections and enrichment space should be used to (1) correct information from the right page (spelling, definitions, etc); (2) add information from the book, computer software, or CD-ROMs; (3) redraw figures; and (4) record concept maps or outlines.

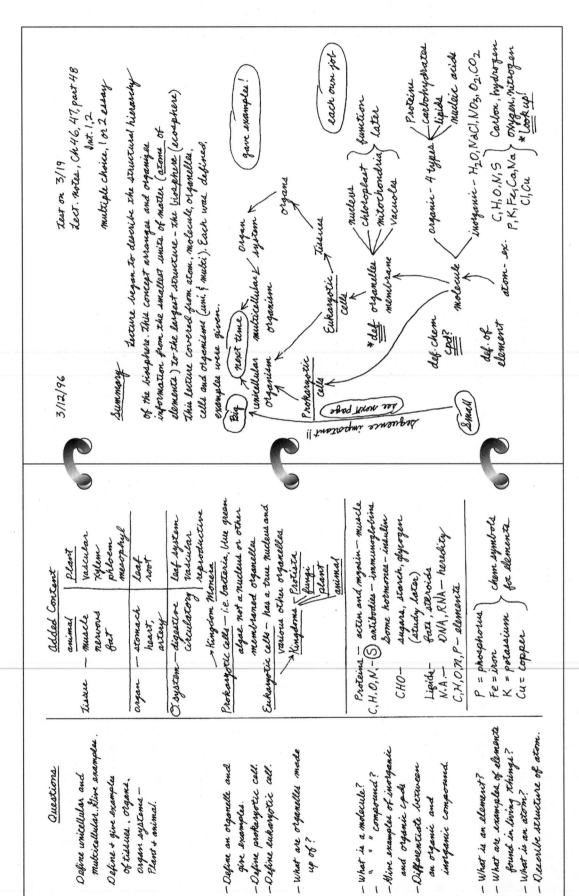

Figure 5.2 Mistakes on the right-hand page should be corrected on the left-hand page. Figures should be clarified or redrawn.

DENNIS THE MENACE

"I GOT A LOT OF ANSWERS, BUT NO ONE EVER ASKS THE RIGHT QUESTIONS."

("DENNIS THE MENACE"® USED BY PERMISSION OF Hank Ketcham and © by North America Syndicate.)

These additions will enrich the notes taken in class. By using the right-page/left-page technique, just about all of the information you will need to learn will be in one place (see figure 5.2).

Question Column The second thing the left page could be used for is to record the questions you create from the information on the right page. In this column you should continually make up and record questions that pertain to the information in your notes.

A question column will (1) help you focus on the material to be studied; (2) enable you to test yourself before the instructor tests you, thus monitoring what you know or don't know; (3) indirectly improve your note-taking and listening skills; (4) make you an active learner by making you a questioning listener; (5) get you to use the information you are hearing and seeing; and (6) help predict what you should know and predict what will be on tests.

Remember, questions should be written in full sentences. Ask your instructor to evaluate them. Ask not only what, but also how, when, where, and why. Avoid asking questions with a yes or no answer. If you do, follow these with "Why" or "Give the reasons for . . .".

Making up many questions will do you no good if you don't use them. During the end-of-the-week review or in your study group, ask and then answer the questions. Cover the notes, read the questions, and recall and recite the answers. Check the notes to see whether your answers are correct. After the first test, compare your questions to the test questions. *How many were helpful? Refine your questioning skill. It will pay off!*

The combination of the notes on the right page and the enrichments and questions on the left page will provide you with a valuable guide to what you must learn. Remember you can modify the format of note-taking to match your style. In the end, you must create a set of good notes to aid your learning and accurately reflect the content of lecture and laboratories.

Using Margins

The margins in your notebook might be used for a number of things:

1. Place a check there if you passed a self-test.
2. Record pages of figures in the text that help answer the questions.
3. Place a question mark to indicate that you should ask your instructor this question.
4. Record the page in your text where the content of a lecture can be found (search of index).
5. List key terms.

Additions to Class Notes

During the end-of-the-week review, you should outline or write a brief summary of the week's work. Construct information charts, tables, or maps. Reorganizing and condensing the week's material will enable you to study the course content in yet another way. These materials should be added to your notebook. By the end of the semester, you will have 14 or 15 packets of weekly reviews that will make studying for the final examination easier.

You might also want to consider adding outlines from specific sections of the book to your notebook. Try to put these next to the class discussion of the same material. Keep similar material together.

The net result of crafting a notebook is that you will create a learning tool. As you do this, you will connect the top part of the learning pyramid with the more fruitful foundation part of the learning pyramid (see figure 3.1).

Review

1. Listening, observing, and writing reinforce each other as you learn.
2. You can improve your listening skills by taking notes, previewing the day's topic, being open to the instructor's teaching style, and being physically and mentally alert.
3. Seeing information written down, such as in your notes, provides visual learning.
4. Lecture and laboratory notes should guide your study; 90 to 100 percent of test material usually comes from class material.
5. You should train yourself to avoid "mental trips." If you miss part of a lecture, leave space in your notes to indicate something is missing.
6. As soon after class as possible, you should correct, clarify, and enrich your notes. Crafting good notes is a study skill.
7. Loose-leaf notebooks allow you to add handouts, returned tests, and your study sheets next to related material in your notes.

8. Formatting your notebook into right-hand/left-hand pages offers major improvements in note-taking. Use the right-hand page to take class notes; use the left-hand page for additions, corrections, and enrichment of class notes. Use the margin to list key terms.

9. Developing a consistent shorthand style can increase your note-taking capacity. Simple symbols such as * and ! can identify important points.

10. Looking at a figure in your text as your instructor explains it will help you remember it.

11. A Correction and Enrichment column in your notebook will provide space for you to add information from your textbook and to make corrections to your class notes.

12. Use a question column to test yourself and to help yourself focus on the material to be learned.

13. The margins of your notebook provide valuable space for listing key terms, page numbers of figures, and page numbers where lecture content can be found.

14. Creating an outline or summary of each week's notes will provide not only a weekly review, but also a study tool to save and use again when preparing for semester finals.

EXERCISE 5

Name _____

Date _____

Listening and Taking Notes

....

1. a. Use this Class Notes Checklist to judge the quality of your class notes. Check "Yes" or "No" to indicate the information included in your class notes. After taking notes for two weeks, you should re-answer this checklist.

Class Notes Checklist

Yes	No	
1 2	1 2	
_____	_____	1. Date and subject of lecture
_____	_____	2. Record of assignments
_____	_____	3. Summary statement of lecture
_____	_____	4. Format
_____	_____	a. consistent style
_____	_____	b. phrases rather than full sentences
_____	_____	c. outline form with indentions
_____	_____	5. Neat and legible
_____	_____	6. Accurate (check with instructor or classmate)
_____	_____	7. Clear, can be remembered
_____	_____	8. Complete
_____	_____	a. record of instructor's hints and gestures
_____	_____	b. record of what is discussed as well as what is written on the board
_____	_____	c. diagrams recorded in large size and appropriately labeled
_____	_____	d. references to diagrams not recorded in notes
_____	_____	e. record of examples of solutions to problems
_____	_____	9. Use of symbols and abbreviations

1 2	1 2	
_____	_____	10. Starring or marking of important points
_____	_____	11. Leaving spaces for points missed or points of confusion
_____	_____	12. Notes taken on the right-hand page
_____	_____	13. Review and clarification of notes within 24 hours of the class
_____	_____	14. Corrections and enrichments added to left-hand page
_____	_____	15. Recording of questions in the Questions column
_____	_____	16. Recording of instructor's questions
_____	_____	17. Comparison of the lecture content with the textbook content and identification of what is covered in both places
_____	_____	18. Outline of pertinent textbook content in Added Content column
_____	_____	19. Summation of week's work in outline, essay, or map

1. b. For each week, record on a separate piece of paper actions to be taken to improve your class notes.

2. How do the recommendations for note-taking in this chapter differ from your note-taking style?

_____

_____

_____

_____

3. If you use pre-written course notes provided by the instructor or a laptop computer for note-taking during lecture, list how you will personalize these notes to record each lecture, express your questions, and review lecture material. Use a separate piece of paper if necessary.

_____

_____

_____

4. Why should you leave blank areas in your notes during the lecture?

_____

_____

_____

5. Can you do a better job of taking notes? List four actions you will take to improve your note-taking.

a. _____

b. _____

c. _____

d. _____

6. Compare notes with a classmate. Discuss any differences. Whose notes are better? Why?

_____

_____

_____

7. How closely do your lecture notes follow the information presented in the textbook?

_____

_____

_____

8. List three reasons to record questions that either you or your instructor ask in a Questions column. (Do you think it's a good idea to use this study skill?)

a. _____

_____

_____

b. _____

_____

_____

c. _____

_____

_____

9. a. Does your instructor use a computer system in class?

_____

9. b. Does your instructor refer to software as a source of information and require the resource to be used?

_____

9. c. If so, how will you incorporate the software information into your study and your notes?

_____

_____

_____

_____

10. Turn to any chapter in your textbook. Create 10 questions from the headings, subheadings, and key terms that appear in the chapter. (You could also use the sample textbook page in figure 8.1 to complete this question.)

_____

_____

_____

_____

_____

_____

_____

_____

_____

_____

_____

CHAPTER 6

Time Management

Objectives

When you have read this chapter, you will be able to answer these questions:

1. What is time management, and how does it apply to my studying?
2. How and why should I set up a semester and weekly schedule?
3. What are some specific ways that I can wisely manage all of my time, especially my study time?

Who hasn't had the feeling that there is so much to do but so little time to do it? Classes to go to, two papers to write, a lab report due, a set of midterm tests coming up, the laundry piling up, the meetings to attend, parents bugging you to write or call, and the list goes on. . . .

Managing Your Time

Time management is an important study skill that can be applied to all aspects of your life. *Schedule specific times to study science, and stick to that schedule. Schedule 6 to 12 hours each week.* If you have a good background in the science you are studying, plus a strong array of prerequisites (see chapter 1), then six to eight hours of study per week might be sufficient to earn the grade you desire. On the other hand, if you have little or no background and only moderate prerequisites, then 12 or more hours per week might be needed to earn a good grade. Study in a variety of ways to learn the information. The study skills inventory in Exercise 3 lists the many things you can do. It is up to you to schedule and direct your study each week. You must tell yourself, "I will study science on Monday from 8:00 to 9:30, Tuesday from 2:00 to 3:15, Wednesday from 8:00 to 9:30, Thursday from 2:00 to 3:15, and Sunday from 8:00 to 10:00." Use the Weekly Time Schedule at the end of this chapter to record times you have committed to classes and to study as well as

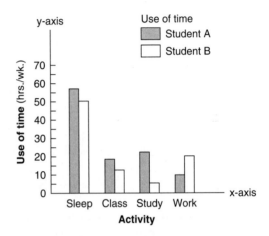

Figure 6.1 Bar graph showing weekly use of time for selected activities. Note that a rate (hours per week) is given for each activity on the y-axis.

to work, meals, travel, or recreation. To help establish your study sessions for science, refer to figure 7.1.

Analyze your use of time by plotting the hours per week spent on various activities on a bar graph (figure 6.1). Note that four different activities are plotted for the two students (A and B). Student A is studying more than B, but B is working more than A. If you don't try to manage your time, you will find it difficult to account for the 168 hours in the week, and many hours will just slip by.

Setting Up a Semester Schedule

After all your classes have met for the first time, set up a **semester schedule.** Using a wall or pocket calendar, indicate during which weeks you have **tests,** when **term papers** are due, and how often **written assignments** must be submitted. This long-term view will chart the *tangible* part of the schoolwork to be accomplished during the semester. Remember, there is a vast amount of study that is *not tangible;* this is the time involved in surveying, questioning, reading, reciting, and reviewing notes and textbooks. The payback for this time is the gradual comprehension of the information and good performance on quizzes and examinations.

Planning a Daily Schedule

After charting the workload for the semester, establish a daily schedule. Make several copies of the Weekly Time Schedule (page 38). Write down on your schedule what activity you will be doing for each day and night of the week. Pay particular attention to scheduling regular study sessions. Be aware that self-discipline and self-direction are vital parts to a successful study schedule.

Keep the following in mind as you plan a study schedule:

A. When?
    1. Schedule study sessions in relatively short blocks of time (40 to 50 minutes) with rewards of 5-to-10-minute breaks. If you can't concentrate for 40 minutes, then study for 20 to 30 minutes without a break, but start to train yourself to concentrate for longer periods of time.
    2. Arrange the schedule so that your science study will be before and soon after the time you attend science class.
    3. Schedule more study time at the beginning of the week than at the end. There is a tendency to goof off toward the end of the week.
    4. Use part of a free day or afternoon to study. The "dead hours" between classes are good times to study. Don't kill these dead hours in the student center.
    5. Plan some time on Friday or Sunday for your end-of-the-week review. This type of study session will enable you to pull together all of the things you studied the previous week. It will also let you preview the coming week. Previewing and reviewing are important study activities.
B. Where?
    1. Keep the study place free from distractions. Train friends and family not to disturb you, or study where they can't get to you. Don't try to study in the cafeteria; studying there doesn't work! It just leads to **R**apid **E**ye **M**ovement—movement of eye from book to passing figures and friends. This form of **REM** generally results in wasted time, frustration, eye strain, and possibly headaches. Use the cafeteria or student center for planned social breaks.
C. Why?
    1. Take control of your time.
    2. Schedule enough time to accomplish the task when you are alert and willing to study.

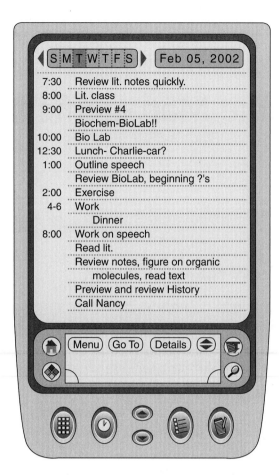

Figure 6.2 An example of a "to do" list. Simple to do, but seldom done.

    3. Set specific goals or objectives for the various blocks of time.
D. What else?
    1. TV and cyberspace are the big "sponges" of time. Avoid being suckered into watching these unless it is on your schedule. Don't try to study and watch TV at the same time.
    2. Plan time to socialize, goof off, and exercise. These activities will add to your overall well-being.

Making a "To-Do" List

You can help keep yourself on schedule by making a "to do" list. You may use an assignment pad, a pocket calendar, or personal computer organizer for this. An advantage of a list is that it helps you simplify your tasks. The list gives you something methodical to do. The tough part is establishing the skill of composing the "to do" list every day (figure 6.2). Resolve to complete the list in the specified time period. Then reward yourself by crossing out the items on the list.

Most people put off doing certain things because the activities seem overwhelming. Easy things, much lower in priority, are done first. Realize that you did the simple things because it felt good to accomplish them. That is the answer. Take the difficult task of studying what seems like massive amounts of scientific information and divide it into a series of small tasks. Follow a schedule and an organized system of study and you will effectively learn and remember subject matter. The alternative is to procrastinate, let it pile up, and try to learn it all in one cram session.

You will be expected to learn the science content presented in lectures and laboratories. The majority of your tests (90% to 100%) are based on these sources of information. There is more content in a textbook than you will study in the course. The lecture and lab content is your guide. If you comprehend that information, you will succeed in the course.

Staying on Top of It

Former students would advise you to "Stay on top of it . . . don't get behind." For every hour you spend in class, spend at least one and one-half to two hours in concentrated study outside class. This time guide presumes you have the prerequisites discussed in chapter 1 and that your grade goal is a B+ or A.

Study involves identifying what must be learned, developing an organized plan to learn, and actually doing the work of learning. After studying is completed, you must review the information learned and practice using the information.

Time Management and Knowing Study Skills

An important time manager is knowledge: knowledge of skills and knowledge of information. If you use the study skill of placing page or figure references in the margins of your notes, you will not have to take the time to look up these references in the index again and again (see chapter 5,

Listening and Taking Notes). As you get to know and apply a reading comprehension technique, you will not struggle awkwardly with what you read or with what was in the figure. To the point, knowing how to learn by using a variety of study skills and activities will make your learning more effective and efficient. Having and using a "learning toolbox" full of study skills will allow you to do a better job of learning in a manageable amount of time. All chapters of this book contain hints and guidance on how to be a more effective learner in science. Use these chapters to fill your "learning toolbox" with tools that work for you.

Computer Software

Computers can help you organize your life. They may help you create schedules and keep track of due dates for assignments and dates for tests. Examples of computerized schedule assistants include palmtop personal organizers, personal organizers available for free on the Internet, and online course management systems (for example, "Blackboard" system) that your instructor may use for the class. Some systems may also include keeping track of money and the other mechanics of your life. These systems might be helpful, but only if you use them properly. For example, don't let the use of an online personal organizer divert your attentions because you get distracted with e-mail or video-clips whenever you log onto the Internet. If you find that you are distracted easily whenever you try to use a computer for personal organization, get a different method for organizing your life, such as a good academic calendar.

Review

1. Time management is an important study skill.
2. Establishing a daily schedule will help you manage your time.
3. Creating a semester calendar helps recognize what and when things need to be done during the semester.
4. Making "to-do" or assignment lists helps with the daily schedule.

EXERCISE 6

Name _____

Date _____

Time Management

· ·

1. a. Does your course use an online course management system? (If you are not sure, ask your instructor.)

_____

1. b. If yes, what is the system called and does it provide a personal organizing system for you to use?

_____

1. c. If no, what personal organizing systems are available to you?

_____

1. d. Which system do you think will work best for you?

_____

2. a. Make a copy of the Weekly Time Schedule located on p. 38. Establish a schedule of your daily activities during the first week of classes. Your schedule is bound to change as you settle into the semester. Use the second copy of the Weekly Time Schedule to record this modified time-management plan.

2. b. Enter this time schedule into the personal organizing system of your choice.

3. a. After your weekly course schedule has been established, list the hours you plan to spend sleeping, attending classes, studying and working.

Time for:

sleep _____   class _____
study _____   work _____
other _____

This is an exercise that is quantitative and will help you to analyze your use of time. You'll be coming up with numbers of hours rather than saying "I studied for a while."

3. b. Plot the results on the bar graph in figure 6.1.

3. c. How many hours are spent each week doing things that do not fall into these four categories ("other")?

_____

Refer to figure 6.1:

3. d. How many hours per week did student A spend in class?

_____

3. e. How much more time did student A study than student B?

_____

3. f. Which student had more unaccounted-for time during the week?

_____

3. g. To which of the two students is your schedule more similar?

_____

4. How do you feel about the recommendation of studying 6 to 12 hours each week?

_____

_____

5. Where do you prefer to study? What aspect of this place detracts from effective studying?

_____

_____

6. Prepare a "to-do" list for tomorrow. Be sure to include specific tasks you want to complete.

_____

_____

_____

7. This is a review type question. List as many study skill activities as you can. Check your list with the list in Exercise 3. How much time would you predict each study activity to take during a study session?

_____

_____

_____

_____

_____

_____

_____

_____

_____

8. On a separate sheet, keep track of how you spend your time for two days. Compare this accounting with your schedule of time. Reevaluate your schedule and time management.

Record times for classes, study, work, recreation, meals, travel, etc. Have you planned 6 to 12 hours of study for science?

Weekly Time Schedule

	Sunday	Monday	Tuesday	Wednesday	Thursday	Friday	Saturday
7:00 AM							
8:00 AM							
9:00 AM							
10:00 AM							
11:00 AM							
12:00 noon							
1:00 PM							
2:00 PM							
3:00 PM							
4:00 PM							
5:00 PM							
6:00 PM							
7:00 PM							
8:00 PM							
9:00 PM							
10:00 PM							
11:00 PM							
12:00 PM							
12:00 midnight							

CHAPTER 7

Study Sessions

Objectives

When you finish reading this chapter, you will be able to answer these questions:

1. How will I know what is important to study?
2. How should I prepare before going to my lecture class?
3. How should I prepare before going to my laboratory class?
4. How should I prepare before attending a recitation session?
5. How and what should I study after attending lecture or laboratory classes?
6. What is meant by an "end-of-the-week" review?
7. How can I set up a group study session, and how will that help me succeed in my science course?

It is easy to say "Know what to study and then learn the stuff." That is just what you have to do. Your class work will be the guide to what has to be learned, and frequent study sessions using various study skills and activities will allow you to learn the "stuff."

During the study sessions, you should:

- Identify and locate information to be studied.
- Organize the information to be studied.
- Reword, rework, and envision the information to interpret the scientific language and figures.
- Practice using and applying the information.

What Is Important to Study?

Here is a guide to help identify the important things to study.

1. Anything mentioned in lecture or lab has a very high priority.

2. If something is mentioned in lecture, lab, and the textbook, it is very important material to learn.
3. The more time that is allocated to a topic, the more time you should spend studying it.
4. You should be able to define scientific terms or symbols used in a lecture or lab and use them in a conversation.
5. Learn the content of figures used to illustrate concepts, principles, processes, and facts.
6. Hints and gestures from instructors highlight important information.
7. Rhetorical questions your instructor asks are clues to what you should learn.
8. Specific assignments indicate which material is important.
9. Problems solved in lecture, lab, or recitation are samples of what you are expected to be able to solve. Practice solving similar problems.

In addition, when studying laboratory material you should know the following:

1. names and functions of tools and apparatus used;
2. names and functions of materials used;
3. results of experiments;
4. units of measurements and symbols of terms and concepts;
5. how to interpret maps, graphs, instrument readings, and results of test reagents;
6. how to describe and discuss what relationships you verified in each exercise or experiment;
7. how to gather information, record results, and take complete notes; and
8. how to correlate terms, units, reagents, materials, and concepts.

Remember, the questions found at the end-of-the-laboratory exercise will test your ability to correlate information and experimentation. Answer the questions.

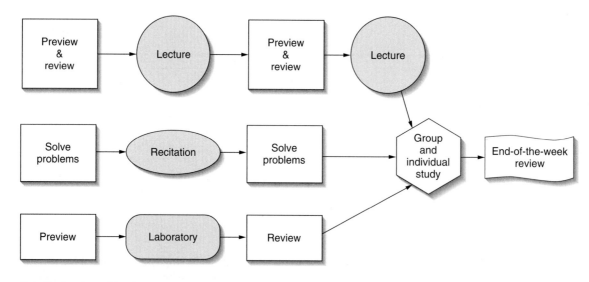

Figure 7.1 Cycle of weekly classes and study sessions. The beginning of the week is on the left, the end is on the right. (The body of this figure is an example of an information flowchart.)

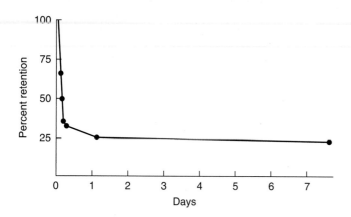

Figure 7.2 Curve of forgetting. Study sessions should take place within 24 hours to help relearn what was forgotten so quickly. (Source: Data from L.F. Annis, *Study Techniques*, 1983, Times Mirror Higher Education Group, Inc., Dubuque, IA.)

Study Sessions

Before going to a lecture class, you should plan and organize a **preview** and **review** of the material. You must prepare ahead of time for each laboratory, and if your science course involves a recitation session, you should complete the assignments before attending the recitation. At the end of the week, a review of the whole week should be done. A study group might come before or after an end-of-the-week review. The schedule of study that should be incorporated into your weekly calendar is shown in figure 7.1.

The need for frequent previews and reviews is based on the tendency for us to forget things quickly. You will forget about 60 to 70 percent of the material you hear in lecture within 24 hours (figure 7.2). If you want to remember it, you must review it frequently.

Set Study Objectives

As you begin each study session, establish **study objectives.** By doing this, you will organize your study and manage your time. The skill of establishing study objectives will, in fact, be another activity to learn the information. You will have to recall, evaluate, and synthesize information into your own organizational scheme of study. Examples of study objectives for four different science topics are as follows:

Biology
- Learn the structure and function of cell parts (organelles) mentioned in lecture and lab.
- Compare prokaryotic and eukaryotic cell structure.
- Make flash cards and redraw figures of cells.

Chemistry
- Review notes on moles and molarity.
- Review sample problems from notes and text.
- Practice answering end-of-chapter questions.

Anatomy and Physiology
- Make flash cards for origins and insertions of muscles covered in lab.
- Do the exercises in the manual and study guide that cover origins and insertions.

Geology
- Construct an information table on minerals studied in lab.
- Make flash cards that include names and compounds, and draw diagrams of mineral crystals.
- Compare photos in text with lab information on minerals.

Preview for Lecture

If the instructor has given definite chapter reading assignments, you can preview the lecture material. The objective of the preview is to develop a general idea about the content of the next class. Don't feel you have to learn it all during the preview. All you have to do is:

- survey the chapter objectives, headings, subheadings, and highlighted words;
- create a "word inventory" of the chapter by recording all the boldface words (don't worry about definitions during the preview);
- read just the introduction and chapter summary;
- survey the figures in the chapter;
- generate questions about the previewed material and read the related questions at the end of the chapter;
- try to predict what will be covered in class;
- if notes are available on the course website, then print and preview these notes.

Review of Lecture

If you get to class early, skim through your notes from the last class. Compare your notes with those of a classmate. Talk about the subject matter for a few moments rather than talking about things not related to class. Recall and recite the material you previewed. Listen to the instructor's review. Put asterisks or exclamation points next to the things in your notes that the instructor mentions. Also, listen for the reviews and previews the instructor gives near the end of class. Take notes on what is said. They will help guide your study.

Long-term memory increases when you review material frequently. A remembering curve with frequent reviews would look something like the one in figure 7.3. If you don't use it, you'll lose it. (Refer to figure 7.2.) Early review of notes will enable you to identify difficult areas. You will then have time to work out the answers or seek help from your instructor or science tutor.

The following activity list is long but important. The bulk of your 6 to 12 hours of study should incorporate many of the following:

1. Review the lecture notes within 24 hours.
2. Highlight or underline key terms.
3. Note gaps. Fill these in or make a note to clarify information with the instructor.
4. Write a brief summary in the space at the beginning or end of the day's lecture notes.
5. Analyze any diagrams in your notes. Compare them to ones that appear in the text. Make any corrections or additions. (Don't forget the right-page/left-page format.)
6. Redraw figures on the opposing blank left-hand page.

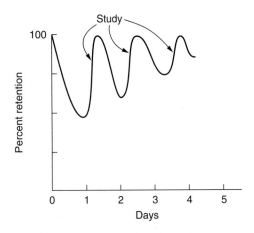

Figure 7.3 Frequent reviews help you retain information. The amount you remember increases as you review and rework information. "If you use it, you will not lose it."

7. Explain or recite the details of the diagram.
8. Compare the definitions and examples given in lecture with those given in the text or glossary. Make corrections.
9. Compare the terms in your lecture notes to the terms you list in your preview term inventory. Check and learn those found in both places. See Exercise 7 number 1.
10. Make flash cards of difficult terms and concepts.
11. Read pertinent parts of the text (S + Q + 3R + P method is recommended, see chapter 8). Take notes on material that will enhance your learning. Record these on the left-hand page. Remember, you should use your class notes to identify what areas of the text should be read.
12. Identify the material covered in the lecture and the text. If it is found in both places, it is information you must learn. If the instructor mentions something that is not in the text, it is also important to learn.
13. Compose information maps or charts when appropriate (see figure 10.2).
14. Compose questions about the information studied. Begin to construct mock tests similar to what your instructor will give.
15. As you study during the week, begin to think about how you can condense and summarize the information you are learning. By doing this, you will have 14 or 15 summaries at the end of the semester. Summaries include outlines (see the sample outline in chapter 2) and short essays. In addition, information maps, charts, and tables allow you to summarize course content efficiently. Summaries will force you to choose important information and organize it in a different and personally meaningful way. You

will have studied the material you are summarizing and will be providing yourself with a useful tool for reviewing the course content.

By doing these things, you will bridge the base and the top of the learning pyramid. You will be taking the time to rework, use, and reflect upon the information you have heard in class.

Preview for Laboratory

When coming to lab, one student often asks another in a relaxed manner: "Hey, what are we doing in lab today?" It might sound friendly and confident, but it's not so smart. Avoid this negative behavior! A positive approach would be to preview the assigned exercises during a study session. You should become familiar with the terms, concepts, and procedures of the laboratory exercises. As you prepare for lab you should do the following:

1. Survey the entire exercise first. Read the objectives carefully. Get a general idea of what is in the introduction and procedures.
2. Survey any questions in the exercise.
3. Generate your own questions about the objectives and procedures. Answer these along with the questions contained in the exercise after you have completed the lab.
4. After doing this, read the introduction more carefully. Relate the information to the lecture if possible.
5. Next, read through the procedure. Underline important parts. Make **flowcharts** or **diagrams** of the procedures. Label the amounts and types of materials to be used. By doing a thorough preview, the mechanics of the laboratory will be less intimidating. It is important that you visualize what you will be doing. Diagrams you create, no matter how crude, will help you visualize the experiments, and that will help you learn.

If you do this type of preview, you will work and learn more effectively and efficiently in lab. During the lab, be sure to take notes on what you have done, what you have observed, and what your instructor has added.

If you complete the laboratory exercises early, stay in the lab and begin to answer questions about the lab. Your instructor will be there to help you.

Review of Laboratory

Complete the study of your laboratory work within two days after finishing the lab. If you have difficulty with the lab report or question pages, you will have plenty of time to get the answers before the next lab period. Guidance for

writing laboratory reports is given in chapter 12. Do the following to complete the study of laboratory exercises:

1. Review the objectives, introduction, and procedures.
2. Clarify and correct notes or data taken down.
3. Analyze the data and observations.
4. Draw conclusions and answer the questions at the end-of-the-lab exercise. Answer the questions you might have generated. Make a note of any unanswered questions.
5. After this is done, visualize what you did, what you observed, and what your results were.
6. Be able to describe the materials and procedures.
7. Relate the laboratory information to the lecture content.

Preview and Review for Recitation Sessions

Problem-solving classes relate the principles and laws discussed in lecture and laboratory. Sets of problems are assigned and then reviewed. The textbook is generally the source of these questions. Chemistry and physics books give sample solutions to the different types of problems presented in the chapter. The questions at the end of the chapter are generally arranged in the same sequence that the material is presented in the chapter. It is vital that you practice solving problems on your own. No pain, no gain. Do the assigned problems. If you need more practice, do the unassigned problems. Work until you are confident that you can solve the problems related to the topics discussed in lecture. Remember, the lecture is your guide. It is the problem-solving process that helps you apply the principles and laws.

After you attend a recitation class, review the solutions discussed in class. Make sure you have no lingering doubts. If you do, go to the instructor or tutors for help. After completing your review, start the next set of assigned problems. Chapter 13 suggests and discusses problem-solving techniques.

Computers

As you study science, your task is to learn the information presented in the course. Computers and software packages could be resources used to convey and learn scientific information. The reliance on computers varies tremendously. Computers can be used for many activities, including the following:

1. word processing;
2. creating graphs and figures;
3. using CD-ROMs to study;
4. reviewing study guides;
5. accessing the web page of the course you are studying;
6. using an online course management system;

7. performing an online "virtual" laboratory;
8. participating in online study groups;
9. accessing web pages of similar courses at other colleges;
10. using your textbook's website to study;
11. communicating with the instructor of the course via e-mail.

As you use these resources, be sure you don't get sidetracked in cyberspace. Remember you are studying a specific science course and must focus your learning on the information presented that relates to lectures and laboratories. Software may contain many symbols in figures that might be in motion. You should refer to chapters 9, 10, and 11 to learn a system of analysis to ensure you are using the software efficiently and effectively. If your instructor uses or recommends the use of certain software, then by all means pursue that resource to enhance your learning.

Websites of instructors and textbooks might provide review questions or samples of tests. If so, these are important ways to test yourself before the instructor tests you. Other study aids might be present. Scan the websites and use what will help you learn.

E-mail access to the instructor might be available. If so, carefully compose questions or communications. Even though your instructor will receive your message immediately, don't expect a quick answer or reply. Thus, e-mail may be a study and review tool to be used throughout the semester. However, it is unlikely to be available or useful the night before an examination or the due date for a term paper. Follow the guidelines given by your instructor concerning the use and function of e-mail.

Group Study Sessions

An effective study group can help pull together the information presented in class each week. Recall that we learn best by teaching others (see chapter 3). Three to six students form a good working group. The group should meet for at least two hours each week with the intent of studying science, not discussing personal or world problems. Work with people whose goals and objectives are similar to yours; avoid academic parasites. It is important that each member of the group prepare for the session by reviewing, correcting, and clarifying his or her own notes. Questions should be generated and problem-solving attempted.

The group must decide on a sequence of study and stick to it. Certainly you must relax and enjoy one another's efforts, but resist any long-lasting distractions to the group effort. You might plan a break to practice a relaxation exercise, listen to music, or have some refreshments. This is the time to practice what you have learned. Your study group might do the following:

1. Compare notes. One person should recite the notes while the others listen and make additions or clarifications. This will be slow-going at first,

but as the semester progresses and everyone's note-taking skills improve, it will get easier.
2. Analyze any figures from the lecture, text, or manual. Have one person explain the figure.
3. Help each other answer questions from the various sources of questions (see below).
4. Review the textbook to highlight areas covered in lecture or lab.
5. Help each other form information charts or maps.
6. Teach and test each other.
7. Thank one another for academic and moral support.

If you are taking a problem-solving course, be sure to work on problems as a group. One person should explain the steps to a solution. Don't get impatient. Check to make sure that everyone in the group understands the process of the solution. Each person in the group should vocalize some part of the group review.

End-of-the-Week-Review

During your end-of-the-week review, recall and integrate the material presented in class. Summarize the week's work into outlines, concept maps, diagrams, or essays. After you do this, test yourself. How well do you know the material? Make up your own test. The questions for this self-testing can come from a number of sources, including the following:

1. Your notes should have questions recorded during the week of study that either you or your instructor asked.
2. You can make up questions from objectives listed in your textbook or laboratory manual.
3. The text and lab manual have questions at the end of each chapter or exercise.
4. The chapter headings and subheadings can be rephrased into questions.
5. You can purchase a study guide.
6. Students can exchange questions to test each other.
7. Learning centers might have computer-based study guides and tests.

Review

1. The most important things to study are those that the teacher discusses in class, especially if a lot of class time is spent on it and it is also covered in the textbook, or if it is part of the homework assignment.
2. When studying laboratory material, it is important that you learn terminology; tool functions; units of measurement; symbols used;

how to interpret maps, graphs, instrument readings, and test results; and how to carry out experiments and record results.

3. Because you will typically forget 60 to 70 percent of the material you hear in lecture within 24 hours, you need to review it frequently to learn it and get it into your long-term memory.

4. Establishing study objectives will help you to organize your study and manage your time.

5. Previewing material for lecture, lab, and recitation classes will help you develop general ideas about the content of the classes, help you generate questions that can be answered during each class, and help you become familiar with the procedures of the laboratory exercise. All of these will, in turn, help you learn more effectively and efficiently during class.

6. Early review of notes allows time for you to correct and enrich notes, write summaries, analyze figures, compare notes to textbook information, and identify information that's important to learn.

7. It is important to complete all assigned problems to make sure that you understand all aspects of the principles and laws being discussed.

8. Computer-based resources can provide course information and possibly be used as an effective and efficient study tool.

9. An effective study group will consist of three to six students who contribute equally to helping one another learn the material.

EXERCISE 7

Study Sessions

....

1. Create a "term inventory" for a chapter in your science textbook. Turn to chapter 2 or 3 in the textbook, note the time, and list the key terms in headings, subheadings, or boldface on a piece of paper as fast as you can. Continue to the end of the chapter. Note the time.

1. a. How long did this "term inventory" exercise take?

   _____

1. b. How many key words are in your inventory?

   _____

1. c. Predict which of the terms in the inventory will be discussed in class. (When you cover this chapter in class, see how accurate your prediction was.)

1. d. Create questions about the first five terms from the inventory.

   _____

   _____

   _____

   _____

   _____

   _____

   Your effort allows you to determine how much time it takes to develop a "term inventory" by using this study skill. It is really easy to create a "term inventory" if you use this skill from your "learning toolbox." The questions you create can become learning objectives.

2. How would you go about finding a figure in your textbook that relates to a figure your instructor created on the board in class?

   _____

   _____

3. Check which of the following people would enable you to fill in the gaps left in your lecture notes.

   _____ instructor   _____ study group
   _____ classmate   _____ science tutor
   _____ lab partner

4. Without looking back, list the different kinds of study activities that you could do in reviewing lecture information.

   _____

   _____

   _____

   _____

   _____

4. a. How many activities did you list of the 15 mentioned in this chapter?

   _____

4. b. Does figure 7.2 accurately represent your retention time for remembering the information? If so, why? If not, why not?

   _____

   _____

   _____

4. c. Make a study activity card to keep at your desk to remind yourself what to do during the review sessions. You can do this by copying the list of study activities suggested in this chapter onto a $5'' \times 7''$ card.

5. Imagine that your instructor does not give specific reading assignments. You have your class notes. How would you go about using your textbook

to confirm and reinforce the information in your notes?

_____

_____

_____

_____

_____

_____

_____

6. Why spend time generating your own questions?

_____

_____

_____

_____

_____

7. Are the graphs in figures 7.2 and 7.3 quantitative or qualitative? State reasons for your answer.

_____

_____

_____

_____

8. A test is in two weeks. Construct a study schedule for the next two weeks and list the activities you will use to prepare for the test. Use a separate sheet if necessary.

_____

_____

_____

_____

_____

_____

_____

9. With a classmate, make a diagram for the following laboratory procedure (protocol). Do this by drawing

symbols for test tubes and indicating the types and volumes of solutions to be placed into the test tubes.

<u>Experiment: Enzymatic Hydrolysis of Starch; Effect of pH</u>

Procedure:

a. Place 1.0 ml of a 0.05% amylase solution into each of five test tubes. Number the tubes 1–5.

b. Place 1.0 ml of a buffer solution (pH$_4$) into #1, 1.0 ml buffer solution (pH$_6$) into #2, and 1.0 ml buffer solutions (pH 8, 10, 12) into tubes #3, #4, and #5, respectively.

c. Add 3.0 ml of 1% starch solution to each of the five tubes. Mix.

d. After 20 minutes, add 1.0 ml of 90% ethanol to each test tube. Mix.

e. Add five drops of iodine solution to each test tube.

10. Outlining a page, section, or chapter is a study activity to give you an overview and summary of information in the book. Outlining can be done during either a preview or review study session. Here's an exercise to help you develop this outline skill:

Outline a chapter in your textbook by putting the headings, subheadings, and boldface terms in indented outline format. Each heading would be numbered; each subheading would be indented and assigned a letter; each boldfaced term would be numbered and indented under the subheading.

For example:

1. Heading

   a. subheading

      1. boldface term 1

      2. boldface term 2

   b. subheading

      1. boldface term 1

2. Heading
and so on.

11. a. Do you have an online study discussion group available to you? _____

11. b. List three important study activities that you can do with an online group.

_____

_____

11. c. List three important study activities that you cannot do with this online group.

_____

_____

11. d. Do you prefer an online study group or a study group that meets in person? Why?

_____

_____

11. e. If you prefer the idea of an online study discussion group, describe how you will make up for the lack of those study activities that you listed in (c).

_____

_____

12. Complete the following table by checking the presence of a study aid found in your science course resource. For instance, the example given in row 1 indicates that key words for study can be found in lecture/lab handouts, final (revised and enriched) notes, the web page, and the textbook, but not on reserve in the library, the CD-ROM, or other software. After completion, this exercise will provide a reference to use while studying. Answers to complete this table will come from the course syllabus, from the lecture announcements, and by asking your instructor.

Study Aid	**Science Course Resources**					
	Lecture/Lab Handouts	Course Notes	Computer Software	Web Page	Textbook	Library
Example: Key words	✓	✓		✓	✓	
Key words						
Review questions						
Sample tests						
Sample problems						
Sample lab reports						
Summary of concepts						

Use of Textbooks

Objectives

When you have read this chapter, you will be able to answer these questions:

1. How can I use my textbook to reinforce lecture and laboratory content?
2. What objective(s) should I have when I begin to read?
3. What reading technique should I use to help me understand what is written in textbooks and laboratory manuals?
4. What would I say if a friend asked me, "What is S + Q + 3R + P?"

The lecture notes, textbook, and laboratory exercises are sources of information you will study. The textbook and laboratory manual are generally used to clarify, reinforce, and supplement the material your instructor covered in class. Follow the instructor's lead as to the depth and breadth of the course content. Compare the reading material to the content of your lecture notes. Concentrate on the information mentioned in both places—that's the important "stuff"!

Computers are allowing more instructors to post class notes, sample exercises, and tests on course websites. These are efforts by the instructor to help the student learn the material of the course. The study skills needed to read textbooks also apply to information available via the computer. You have to decide how this information is organized, how it relates to your learning style, and how the web information will help you learn the information to pass the tests.

Textbook Organization

Believe it or not, a textbook is organized to help students learn. The information is challenging, but publishers make

Table 8.1 Textbook and Software Use

Textbook Feature	Type of Help
Chapter introduction	Overview
Heading	Identify
Subheadings	Identify
Boldface	Identify
Text	Explain
In-text summary	Summarize
Figures	Visualize, summarize
Table	Identify, summarize
Boxes	Apply, relate
Sample solution	Model
Chapter summary	Overview, summarize
Chapter questions	Test
Glossary	Define
Index	"Finder"
Appendix	Special information
Software (computer)	
CD-Roms	Visualize, explain, summarize, test
Videodiscs	Visualize, summarize
Textbook website	Visualize, explain, summarize, test

an effort to break the subject matter into manageable segments. Table 8.1 lists textbook features and indicates the type of help they provide.

Compare your science textbook with this list. Does it contain the same elements? A checklist in the exercise at the end of this chapter covers these elements and more that may be part of your textbook. Figure 8.1 is a reduced page from a general biology textbook. Note the aids it offers to help you identify things to learn.

Some molecules that occur in organisms are simple organic molecules, often with a single reactive functional group protruding from a carbon chain. Other molecules, called **macromolecules,** are far larger and often play a structural role in organisms or store information. Most of these macromolecules are themselves composed of simpler components, just as a wall is composed of individual bricks. The similar components that are linked together to form a macromolecule are called subunits. Macromolecules fall into the following four classes: carbohydrates, lipids, proteins, and nucleic acids (table 2.2).

The molecules formed by living organisms all contain carbon and are called organic molecules. Large organic molecules, or macromolecules, play a structural role in organisms or store information.

Table 2.2   Macromolecules

Macromolecule	Subunit	Function
Carbohydrates		
Glucose (monosaccharide)	—	Energy storage
Starch, glycogen (polysaccharides)	Glucose	Energy storage
Cellulose (polysaccharide)	Glucose	Component of plant cell walls
Chitin (polysaccharide)	Modified glucose	Cell walls of fungi; outer skeleton of insects and related groups
Lipids		
Fats	Glycerol + three fatty acids	Energy storage
Phospholipids	Glycerol + two fatty acids + phosphate	Component of cell membranes
Steroids	Four carbon rings	Message transmission (hormones)
Terpenes	Long carbon chains	Pigments in photosynthesis
Proteins		
Globular	Amino acids	Catalysis
Structural	Amino acids	Support and structure
Nucleic Acids		
DNA	Nucleotides	Encoding of hereditary information
RNA	Nucleotides	Blueprint of hereditary information
ATP	Nucleotides	Energy transmission and conversion

Polymers

Many macromolecules are polymers. A **polymer** is a molecule built of a long chain of similar molecules called **monomers,** like railway cars coupled together to form a train. Complex carbohydrates, for example, are polymers of simple monomers called sugars. Enzymes, membrane proteins, and other proteins are polymers of monomers called amino acids. DNA and RNA are two versions of a long-chain molecule called a nucleic acid, which is a polymer composed of a long series of monomers called nucleotides.

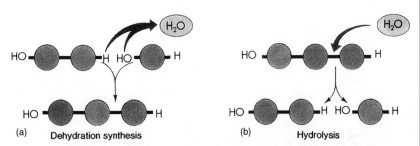

Figure 2.16 Dehydration synthesis and hydrolysis.
(a) *Biological molecules are formed by linking subunits. The covalent bond between subunits is formed in dehydration synthesis, a process during which a water molecule is eliminated. (b) Breaking such a bond requires the addition of a water molecule, a reaction called hydrolysis.*

Making (And Breaking) Macromolecules

Although as shown in table 2.2, the four different kinds of macromolecules are assembled from different kinds of subunits, they all put their subunits together in the same way: A covalent bond is formed between two subunits in which a hydroxyl group (OH) is removed from one subunit and a hydrogen (H) is removed from the other. This process is called **dehydration** (water-losing) **synthesis** because, in effect, the removal of the OH and H groups constitutes removal of a molecule of water (figure 2.16*a*). In the synthesis of a polymer, one water molecule is removed for every link in the chain of monomers. Energy is required to break the chemical bonds when water is extracted from the monomers, and so cells must supply energy to assemble polymers. The process also requires that the two monomers be held close together and that the correct chemical bonds be stressed and broken. This process of positioning and stressing is facilitated by helper molecules that do not themselves change, in a process called **catalysis.** In cells, catalysis is carried out by a special class of proteins called **enzymes.**

Figure 8.1 A typical page in a biology textbook. Note how the page is divided and formatted to help you identify information to learn. (Page from Peter H. Raven and George B. Johnson, *Understanding Biology,* 3rd edition. Copyright © 1995 McGraw-Hill Company, Inc., Dubuque, Iowa. All Rights Reserved. Reprinted by permission.)

Reading Skills

As you study the textbook, you must be able to:

- establish goals before you start to read;
- read flexibly, for details or for general concepts;
- learn from the various types of figures;
- be able to understand the symbols of science;
- read sequences of directions accurately and carefully;
- generate and interpret questions;
- analyze and evaluate data and information;
- draw conclusions based on the analysis;
- apply the information in critical thinking about everyday or scientific problems.

You should keep in mind that science textbooks contain a lot of information on each page. They are information dense. In addition, most textbooks contain more information than is presented in class. Thus, before you begin a reading assignment you should:

1. know where to find the material;
2. set reading objectives; and
3. adopt an organized method of study.

Step 1: Knowing Where to Find the Material

The chapter assignments, the lecture content, and lab exercises will help you locate the material to be studied in your textbook. Key terms recorded in your notes can be looked up in the index. This will give you specific pages to read. Again, if the information is mentioned in the lecture and the text, this is the material you should be sure to learn. It is prime test material!

Step 2: Setting Reading Objectives

You must have an objective in mind when you use the textbook, and you must use a consistent technique to fulfill that objective. Don't just read. Science books are not novels. Create lists of objectives to help you focus on what to study. The following are examples of learning objectives:

- I am going to learn about the function of the pituitary gland.
- I am going to learn about the concept of neutralization.
- I am going to compare metamorphic and igneous rock.
- I am going to compare velocity and acceleration.
- I will distinguish between covalent and ionic bonds.
- I will distinguish between mitosis and meiosis.
- I will compare the content of today's class notes to the content in my textbook.

Step 3: Adopting a Study Method

The third step is to learn the material you are studying. How do you get it out of the book and into your "thick" head? If you're going to try to do some serious studying, to use your time effectively and efficiently, then choose the proper environment. Reduce distractions; don't play your favorite music, don't have a picture of your boyfriend or girlfriend in front of you, don't have the TV on, don't try to study in busy areas of the library or student center or in the cafeteria. Yes, you can study with all these distractions, but it will not be as efficient; various stimuli will be competing to get into your cerebral cortex. See chapters 3, 6, and 7.

Survey, Question, Read, Recite, Review, and Practice (S + Q + 3R + P)

Adopt a concrete, logical method to study your textbooks and to fulfill the learning objectives you have listed. Avoid the predicament of "I studied the text for hours; I can't understand why I got a 52. I must have overstudied. It's not fair. Gary got an 84 and hardly studied." A recommended method is *S*urvey, *Q*uestion, *R*ead, *R*ecite, *R*eview plus *P*ractice, or S + Q + 3R + P.

S + Q + 3R + P is an important study tool. This technique will help you understand the textbook and laboratory manual. In addition, the pattern of the technique can be applied to reviewing your notes and understanding figures that appear in the textbook. Make an S + Q + 3R + P study skill card. Keep this as a bookmarker and a reminder of the sequence of this reading technique.

Survey

A survey is meant to familiarize you with the material. You are essentially warming up to reveal what objective you should accomplish during the Q + 3R segment of studying the text. Read the introduction to the chapter, the list of objectives if present, and just the headings and subheadings. Scan the terms in **boldface** and the **summary.** Do not read the text of the chapter. **Scan** the **captions** and **body** of figures. Develop a general view of the chapter or the section of the chapter you are surveying. Scan the **questions** or **problems** at the end of the chapter. Surveying should be a short burst of concentrated study; 10 to 15 minutes should do it. Remember, a quick overview of the topic is all you want to do.

You can also make a separate "how to survey" card, listing the surveying activities on a 3″ × 5″ card. If you forget what to do when surveying the text, refer to the list. Remind yourself about good surveying techniques until you have it down pat.

Don't start to survey and read the whole chapter if you will only be covering one-third of the chapter in lecture.

Two additional exercises could be done during the survey if you think it worthwhile. "Race" through the chapter to create a **term inventory** of all of the scientific terms in headings, subheadings, and boldface. In 10 to 15 minutes, you should be able to list all of the terms you might have to learn. At a later time, you can check off which terms your instructor uses. Those words are the minimum list of terms to learn. You could also copy the major headings and subheadings onto a piece of paper in an outline format. If you do this, you will create an outline of the chapter. This outline along with the term inventory would list the chapter's content.

Question

After surveying, scan the chapter once again, but this time generate questions from the headings, subheadings, and boldfaced terms. Recite these or record them in the Questions column in your notebook (see chapter 5). Re-read the questions you have generated in your lecture class. You will begin to have an idea of what to study. These questions can serve as learning objectives.

Read

Now you are ready to read the chapter. Note the number of science terms on the sample page (figure 8.1). Does the general reading you have done in your life help you comprehend this type of reading? You should read slowly and carefully. Remember, textbooks are information dense. You might find it helpful to guide yourself through the text with a pencil or to read the difficult parts out loud or in a whisper. Pencil tracking and verbalization seem to make difficult sections seem somewhat easier. You might want to underline or highlight certain phrases and star or mark certain concepts in the margins. Avoid underlining or highlighting entire paragraphs. As you read, find the answers to the questions you have generated and the objectives you have established.

In addition to this careful and detailed (every word counts) study, you should stop every once in a while to cross-reference the information you are reading with the information in your notes. Place some sort of identifying symbol in your notes that indicates the information is in the textbook. This is the information you must learn. How did the instructor teach the material you are reading? All of this takes time and effort. That is why you need to spend 6 to 12 hours of study spread out in a number of study sessions each week. Read and study to answer questions and reinforce the content of lecture and laboratories.

By using a pencil, answering questions, and fulfilling learning objectives, you will be a more active and focused reader.

Recite

Before racing to "finish off the chapter," pause to recite or paraphrase each paragraph in your own words and use the new terms. Answer the questions you created. Recite the equations and explain the relationships represented in them. Try to envision the figures presented or the example problems. Pat yourself on the back if you recited correctly. Go back and re-read if you stumbled or drew a visual blank.

Review

When you finish several sections or a chapter, review the material. Repeat the survey. Go over the headings and the subheadings, key terms, figures, and summary. This is a good time to compare your lecture notes and the textbook material. Note the similarities and differences. By doing this you will be able to clarify and correct your notes. Be sure to make corrections or add enhancements to your notes on the left-hand page of your notebook.

Construct concept maps, tables, charts, or outlines to summarize what you have learned (see chapter 11). If the instructor did not talk about or assign a specific section of the chapter, this should not be part of the learning objectives list. However, if you're interested in that material, go ahead and read it. For instance, during a review of what you have just read about S + Q + 3R + P, you could write an outline, construct a concept map, or diagram a flowchart of S + Q + 3R + P (figure 8.2). This detailed analysis of a skill looks complicated as a flowchart. If you analyzed the process of swinging a baseball bat, it would likewise look complicated. However, once practiced and perfected, it becomes second nature and seemingly simple.

Practice

Practice using the material you have learned. Try to apply the information to the world around you. This will be easier for some courses than for others. Read articles in papers and magazines that relate to your study. Explain what you have learned to classmates, parents, friends, or pets. Draw diagrams about the material and explain them. In a problem-solving course, practice solving assigned and unassigned problems. Problems will surely be on the tests. You can't wish your way through them. Most people will agree that "If you don't use it, you'll lose it." If you use what you've learned, you'll retain it.

Organization of Laboratory Manuals

Most manuals are workbooks in which you are expected to record data, write your analysis, record your conclusions, and answer questions. Compare the organization of the

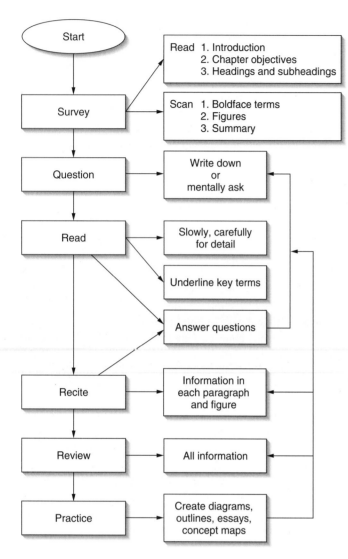

Figure 8.2 A flowchart depicting the sequence of S + Q + 3R + P. Note how the activities of this study skill loop back and reinforce each other.

laboratory manual to the textbook checklist in the exercise at the end of this chapter. You will see similarities and differences. How does the organization of the laboratory manual help you identify the work to be done and the information to be learned in each exercise?

As a general rule, exercises have an introduction, a statement of objectives or problems, a list of materials, a sequence of procedures, and a place to record observations, analyses, and conclusions. Questions about the exercise can be found at the end of the chapter.

Reading Laboratory Exercises

The organized S + Q + 3R + P method of reading textbooks applies to the exercises in laboratory manuals. Use the S + Q + 3R + P method to study the assigned exercises in the laboratory manual. This should be done before you attend

the laboratory class. See chapter 7 for other clues to studying the lab content. It should be emphasized that during the review of the lab procedures, you should create flowcharts, concept maps, or diagrams in preparation for the actual laboratory class.

 # Computer Resources

Software

Computer software is, in a sense, another form of a textbook. Thus the S + Q + 3R + P technique should be applied to use these resources efficiently and effectively. As in any textbook, there will be more information given than presented in lecture or lab. It will be important for you to concentrate on the information that supports lecture and labs. Establishing study objectives (see chapter 7) will help direct your use of the software. In addition, you should take notes about the steps followed to reach valuable parts of the software; this will enable you to relocate the same part of the program at a later date.

Hopefully your instructor will give you some guidance about the software that supplements the course presentation. If instructors don't mention the use of software programs, then they don't feel that this resource will be needed to learn the content of the course. In this case, they may have evaluated the available software and decided it was not effective for the course objectives. On the other hand, they might not have evaluated and integrated available software into the course. Publishers provide CD-ROMs and websites with some textbooks. Instructors might not even mention these and leave it up to the student to determine the usefulness of these resources.

Students have found that software reviewing high school science presents information similar to the college science course. The often similar presentation gives a good foundation to study the information presented in college science courses. In addition, other programs are available to help review basic skills such as writing and mathematics.

Internet: Searching Effectively

The Internet is being used as a resource for study and research. Some risks involved in use of this resource are:

1. searching inefficiently for information;
2. spending too much time searching without fulfilling learning objectives;
3. receiving biased and irreputable information since many web pages are personal opinions and not peer-reviewed publications.

If you plan to use the Internet in your science course, the exercise at the end of the chapter will help you experience focused Internet searches and information compilation. The exercise is not meant to train you on Internet software use. You may have to spend time learning how to use the search engine programs available on your Internet software. Before you start searching for information, you must formulate a question and define a research objective. After this is done, then you can begin exploring the Internet using search terms.

When defining search terms, remember the use of "+" between key words usually tells the search engine to search for both words occurring together. For whatever search engine you use, check the search options, where you are able to select an option for a search for both words. For example, "Mars + Life" will show sites that have "Mars" and "Life" together. If the program searches for any word in your string ("Mars," "Life") you may end up with too many site choices. This type of search would give you any site that has the word "Mars" or "Life" in it. Imagine how many sites have the word "Life" in them!

Internet searches will provide you with too much information. Some of that information is good, and other information is bad. Some websites are backed by reliable organizations, others do not provide accurate information. Questions to ask yourself when you are deciding whether or not a website is reliable and accurate are:

1. Is the site's information developed by a well-recognized organization or expert in the field of science?
2. Why did the site's author put this information on the web? (For example, if there are political reasons for posting information, this site is not likely to be helpful to you for your science class.)

3. Does the site's information agree with information from other resources (such as other websites, your textbook, your instructor)? If you can corroborate information, then it is likely that the site's information is accurate.

Review

1. The lecture and lab content are the guide to the information you should study. If you do not have complete notes, then you will have difficulty knowing what to study in the textbook.
2. Your textbook will contain information and examples to reinforce and supplement the class content.
3. If information appears in lecture or lab notes and in your text or manual, then that is the important information to learn.
4. The reading and laboratory assignments will help you identify what you should study.
5. Before you use your text, you should formulate learning objectives. Use the objectives to guide the reading of the textbook.
6. You can read for understanding by using the technique of $S + Q + 3R + P$.
7. $S + Q + 3R + P$ means *Survey, Question, Read, Recite, Review* plus *Practice*.
8. Skills for reviewing and using textbooks can also be used for computer resources.

EXERCISE 8

Name _____

Date _____

Use of Textbooks

....

1. Check which of the following are included in your textbook.

_____ Preface with hints on how to use the book
_____ Table of contents
_____ Chapter titles
_____ Chapter objectives
_____ Headings and subheadings
_____ Bold printing of key words
_____ Key term list
_____ Introduction to each chapter
_____ Figures within chapters
_____ Marginal comments
_____ Key concepts repeated in boxes or different font
_____ Portions in boldface print
_____ Boxed discussions or examples
_____ Chapter summary
_____ Review questions
_____ Examples of solutions to problems
_____ Use of symbols
_____ Use of mathematical equations
_____ Answer keys
_____ Chapter test
_____ Glossary
_____ Index
_____ Website

2. Which two elements can help you locate a topic in the text?

_____

_____

3. Which one can help you define a specific term?

_____

4. Which three can be used to generate your own questions?

_____ _____ _____

5. Why do certain terms in your text appear in boldface?

_____

_____

6. Most textbooks contain more information than is presented in the course. How can you select the material that you should study in the text?

_____

_____

7. a. What is a learning or reading objective?

_____

_____

7. b. What is the value of establishing learning and reading objectives before you start to study or read?

_____

_____

_____

8. List three learning objectives for reading this chapter.

_____

_____

_____

9. Do you think it is worth your time to list a series of objectives? Why or why not?

_____

_____

10. a. What does S + Q + 3R + P mean?

_____

10. b. Summarize what should be accomplished in each step of S + Q + 3R + P.

_____

_____

_____

_____

_____

_____

_____

11. How does the S + Q + 3R + P method differ from the way you have used textbooks in the past?

_____

_____

_____

_____

12. Refer to figure 8.1. Assume the reading objectives are to learn:

12. a. the characteristics and examples of polymers and monomers.

12. b. about the chemistry of making and breaking macromolecules or polymers.

12. c. about the role of enzymes in dehydration synthesis and hydrolysis.

S + Q + 3R + P this figure. After surveying the page, record questions that must be answered. Then read the page and recite the information. Decide what you would do to review the information. Answer the questions you generated. Compare your work with another student's work.

_____

_____

_____

_____

_____

_____

13. How will you incorporate the study skill of S + Q + 3R + P into your cycle of weekly study?

_____

_____

14. Create a term inventory for the page in figure 8.1.

14. a. How many terms have you listed?

_____

14. b. How much time would be needed to learn the meaning of these terms?

_____

15. How can you use the textbook headings and subheadings to help you learn?

_____

_____

16. Why should you compare the content of your lecture notes to the information you read in your textbook?

_____

_____

17. a. Does a CD-ROM come with your textbook? How do you plan to integrate the CD-ROM information with your class notes and textbook reading?

_____

_____

_____

17. b. Does your textbook have a website? If so, what is the website address, and how do you plan to integrate the information on the website with your class notes and textbook reading?

_____

_____

_____

_____

_____

18. The objective of this exercise is to give you experience in a focused Internet search. The following table defines four questions, states search objectives for each question, and suggests search terms. Choose one or more of the examples to gain practice with Internet searches. Answer questions a–f to help guide your work. Use the below space and any extra pages to record your answers.

18.   a.   Choose one example from the table.

18.   b.   Using the suggested search terms, find two websites from which to gather information to fulfill the search objectives. Write their addresses in the following spaces:

Site 1:

Site 2:

18.   c.   Gather information from each website to reach the search objective(s).

18.   d.   Did each site give you enough information to reach the search objective(s)?

_____

18.   e.   Does the information gathered from both sites agree or disagree?

_____

18.   f.   List three actions you can take to verify the reliability and accuracy of your collected information.

_____

_____

_____

Question(s)	Search Objective(s)	Search Term(s)
What evidence do we have that there is life on Mars?	Search for information that gives examples of data collected to show life or the possibility of life on Mars.	Mars + Life
Is global warming a real threat?	Search for evidence to support that global warming is occurring and to support that global warming is not occurring.	Global + Warming
What is osteoporosis? Do only women get it? How can it be prevented?	Search for information on osteoporosis, especially a good definition. Try to find percentages of women and men who have it and recommendations on how to prevent it.	Osteoporosis
What is El Niño, and why does it happen?	Search for a simple description of El Niño and its cause.	El Niño

CHAPTER 9

Terms, Symbols, and Figures

Objectives

When you have read this chapter, you will be able to answer these questions:

1. What terms must I learn in the science course?
2. How in the world can I learn all those specialized terms?
3. What do the symbols of science mean?
4. How can I learn the symbols used in science?
5. Why is it important to learn to recognize and interpret symbols?
6. Computers and the great "action" programmed into them will make learning fun and easier . . . won't it?
7. What are the different types of figures used in lectures, textbooks, and laboratory manuals?

Scientists have investigated objects as small as the parts of the atom and as vast as the galaxies of the universe. Astronomy, biology, chemistry, ecology, meteorology, and physics have each developed a unique body of knowledge. Special terms name structures and label processes. A ribosome is the name of a cell structure. Dehydration synthesis and translation are processes that occur in the ribosome to produce protein polymers. A number of different molecules participate in a series of chemical reactions. An explanation of the step-wise sequence of molecular reactions would describe the mechanism of protein synthesis. Terms and their abbreviations (a form of symbol) identify the measurements of distance, time, energy, and mass.

Symbols are used in science to represent the structures and processes. Abbreviations are just one form of a symbol. Letters, shapes, colors, arrows, and diagrams are other common symbols that represent either structures or processes. Terms and symbols are combined into figures that display scientific information. Science textbooks and computer software use different types of figures to convey information.

Learning Terms

Knowing the vocabulary of science is essential to comprehend and communicate scientific information. As you study science, you must learn the terms, measures, expressions of rates, and whatever else is used. If your instructor uses it, then you must learn it. New terms are generally introduced in rapid fire (without mercy, it might seem). Many of the terms are difficult to spell and even more difficult to pronounce. You have to learn the correct:

- meaning;
- usage;
- spelling; and
- pronunciation.

Descriptive courses like anatomy might introduce 40 to 60 new terms in one lesson. Problem-solving courses introduce fewer terms but apply basic concepts and relationships (rates and ratios) to different conditions or problems.

If you memorize the meaning of the terms, you will have the basic tools to communicate, but not necessarily an understanding of the subject. The next challenge is to integrate these words and symbols into concepts, as expressed in complete sentences, paragraphs, solutions to problems, or complete thoughts. This comprehension will enable you to answer questions, evaluate information, solve problems, and communicate in science.

Some textbooks help teach terms by giving their pronunciations; for instance, "sternocleidomastoid" might also be given as (ster″ no-kli″ do-mas″ toid). As you begin to learn a complex word like this, you should try to figure out its "roots." Why are "sterno," "cleido," and "mastoid" compounded into one word? Terms have root words that have a meaning. For instance "port" means "to carry." By adding the prefix "ex" (meaning "out"), you construct the word "export" meaning "to carry out." The suffix "graph" means an "instrument to record." It can be combined with the prefix "electr" and the word root "cardi" to form the word "electrocardiograph." What does the word mean? The prefix "photo," of

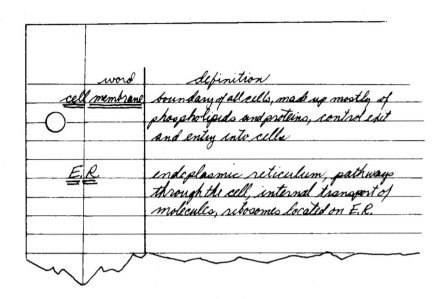

Figure 9.1 Example of a vocabulary-definition list. Note how the spacing helps focus on the terms and definitions.

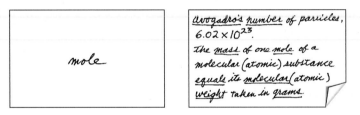

Figure 9.2 An example of the two sides of a flash card. One side has a single term, the other side has information about that term.

photosynthesis or phototropism, refers to light. The suffix "lysis" means "to break up." The terms of analysis and hydrolysis use this suffix. If you know the meanings of prefixes, roots, and suffixes of words, you can begin to figure out the meanings of words on your own. You should try to develop an understanding for the words you are learning (memorizing).

Dictionary

The instructor or your textbook might use words not specific to science that are not part of your vocabulary. Note these words in the margins of your notebook. Use a dictionary and your textbook's glossary to clarify the meaning and usage of these words. Study their roots and any prefixes or suffixes. Begin to use these words to add them to your vocabulary.

Part of your study sessions should be devoted to systematically identifying and learning the terms. These are two important study skills. You can identify terms to learn by creating a list from:

- lecture notes;
- lab notes;
- term inventory of textbook;
- term inventory of lab exercises;
- end-of-chapter key terms list; and
- instructor's vocabulary list.

Two ways of learning terms are to maintain a running vocabulary definition list or to establish a flash card system. Recall the graphs for forgetting and learning (figures 7.2 and 7.3).

Vocabulary-Definition List

The term should be recorded in a column on the left-hand side of a page; the definition is recorded next to the term on the right-hand side (figure 9.1). You can refer to the glossary or your notes before expressing the meaning in your own words. Review the list frequently. To test your knowledge, cover the meaning and recite the meaning. Of course you could also cover the term, read the definition, and recall the term. Use the terms or short phrases in sentences and relate them to other terms to better understand and remember them. Place a check mark next to the terms you have learned.

Flash Cards

Use flashcards to test yourself often. Write one term or short phrase on one side of the card. (Consider cutting a page into four pieces rather than buying cards. It's cheaper.) On the other side, record the meaning and related information (figure 9.2); for instance, a cell membrane card might include the information given in figure 9.1 and references to osmosis, active transport, endocytosis, and exocytosis. Keep the flash cards short and to the point. If need be, make another card with the same term but include simple line diagrams. Resist just copying notes onto the card. Carry your stack of cards with you. Whenever you have a few free moments, review the cards. Begin to separate the stack into "known" and "not known" piles. Continue to work on the "not known" pile, but review the "known" stack every once in a while.

Learning Symbols

Something that represents something else can be considered a symbol. Symbols are the abbreviations, arrows, letters, or special shapes that represent a term, structure, or process. A variety of symbols might be linked together in a figure to represent the many steps of a mechanism. The understanding of individual symbols is necessary to comprehend the figures that are included in the textbook or used by the instructor. Identifying and learning symbols will be another challenge. This chapter will help you focus on learning how to identify and learn symbols. Chapters 10 and 11 will help you develop the skill of interpreting figures that use different types of symbols.

Symbols are included in the text material, or they might be part of a figure or table. For instance, the symbol for a hydroxyl group (OH) is used in the text of figure 8.1. Other symbols in the reproduced textbook page are various colored shapes that are creations of artists and authors. A water molecule is represented by the symbol H_2O in an oval shape. Circles represent subunits or monomers. The dark lines connecting the circles are symbols representing covalent bonds. What are two other symbols used in figure 8.1?

Take special note of the way your instructor uses symbols. Letters, numbers, different shapes and shading, specialized symbols, and arrows are all important in the communication of information. The assumption is made that you will be able to understand and communicate science in either written or symbolic format.

Many symbols must be memorized. Again, you may use either a symbol-definition list or flash card system to help you do this. Other symbols that are in books or AV materials must just be "figured out" or interpreted; for example, the body of the illustration shown on the sample book page (figure 8.1) does not specifically identify the circular symbol as a subunit. You must read the caption to find that out. In addition, you must relate the information in the table to the figure to identify a specific type of subunit or monomer that is represented by the circles.

This skill of identifying and learning symbols can be accomplished in two ways. First, read the caption and the textbook explanations of illustrations. Take time to relate the terms or words to the symbols in the figure. In addition, you can learn to interpret symbols by listening to and observing how your instructor uses the symbols to communicate the scientific information. Carefully copy the symbols your instructor uses into your notes. Label what these symbols represent. If a symbol appears in your notes but makes no sense to you, find out what it means.

Table 9.1 Common Measurements in Science

Symbol	Term Symbol Represents
m	meter
cm	centimeter
mm	millimeter
μ	micron or micrometer
nm	nanometer
Å	Angstrom
°C	degree centigrade or Celsius
°F	degree Fahrenheit
K	Kelvin
kg	kilogram
g	gram
mg	milligram
L	liter
ml	milliliter
cc, cm^3	cubic centimeter
V	volt
W	watt
cal	calorie
J	Joule

Standardized Symbols

Examples of symbols used in science are given in tables 9.1, 9.2, and 9.3. These tables are by no means a complete list of important symbols. Remember, if your instructor uses a symbol, you must learn it and be able to use it. Note that table 9.1 gives examples of units of measurements. When different measurements are related in math-based problems, units of rate result. Meters per second, heartbeats per minute, moles per liter, and calories per gram are examples of rates. The rate meters per second can be shortened to the symbol m/sec, and moles per liter can be expressed as mol/L. Units and rates are particularly important when you solve math-based problems. Don't lose track of what units are given in the problems, and don't forget to label the units or rates for the answers you have calculated.

Symbols to Be Interpreted

In many figures, a symbol has been created by an artist or author that represents something. These kinds of symbols are not standardized symbols, but you should be able to interpret what they represent. Four examples are found in figure 9.3.

Table 9.2 Abbreviations Used as Symbols in Science

Science	Symbol	Meaning
Biology		
	DNA	Deoxyribonucleic acid
	F_1	First filial
	P_p	Heterozygous genotype
	XX	Two female sex chromosomes
	ADH	Antidiuretic hormone
	HIV	Human immunodeficiency virus
Chemistry		
	O	Oxygen and one atom of the element
	Na	Sodium and one atom of the element
	Cl	Chlorine and one atom of the element
	pH	Potential hydrogen negative logarithm of concentrations of hydrogen ions
	s,p,d,f	Subshell of orbit
Physics		
	F	Force
	m	Mass
	d	Density
	P	Pressure
	V	Volume
	A	Area
Meteorology		
	Ci	Cirrus cloud
	St	Stratus
	Cu	Cumulus cloud

Table 9.3 Examples of Other Symbols Used in Science

Symbol	Meaning
♂	Male
♀	Female
[ ]	Concentration
%	Percent
<	Less than
>	Greater than
Δ	Change or difference between
+	Positive charge, add, plus
−	Negative charge, subtract, minus
O=O	Two atoms of oxygen with a double covalent chemical bond
H_2O	A water molecule, two hydrogen atoms, one oxygen atom
U_{235}	The element of uranium, one atom of the element, 235 is its mass number
$7O_2(g)$	Seven molecules of oxygen gas
$C_6H_{12}O_6$	Represents one glucose molecule; 6 carbon, 12 hydrogen, and 6 oxygen atoms

1. (a): Note a cell is symbolized with a nucleus (the dark oval shape), the cell membrane (the line of the irregular oval shape), and the Y-shaped molecules attached to the membrane.
2. (b): The areas with blocks, dashed lines, dark big dots, small dots, and the black and grey parallel lines represent layers of different rocks.
3. (c): The shaded areas represent the areas in the orbits of an atom where you could expect to find the rapidly moving electrons.
4. (d): The three shapes represent different molecules as indicated by the labels. The different shapes indicate how they could all fit together.

What Can Arrows Mean?

Arrows are important symbols. They are used frequently and can have many different meanings. You must be able to **locate** **arrows** and **interpret** their **meaning.** Arrows might simply point out something. Other times the arrow indicates a figure has been magnified and is pointing to magnified subject.

Arrows can also represent a process carried on by structures. A process is an action or a series of actions directed to some end result. Arrows are used to symbolize the process and the direction of the change. Other symbols in the figure represent the structures involved in the action. Differently shaped arrows might be used in one figure, as is done on the illustration shown on the sample page (figure 8.1). It is important to be able to interpret the process the arrow represents, the direction of the process, and where the process begins and ends.

Examples of the use of arrows are shown in figure 9.4. In (a), note the two arrows used to point to the proton or neutron. [What do the (+) and the (−) symbolize? What do the oval lines represent?] The curved arrows in (b) represent the steps in the processes of lithification, metamorphism, melting, cooling, and weathering. The three straight arrows indicate alternate pathways for rocks to undergo changes in the rock cycle. In (c), the arrows are labelled in this representation. The text indicates that DNA is a template for the synthesis of mRNA rather than DNA turning into RNA.

Can You Find the Terms and Symbols?

Figure 9.5 contains eight terms as well as 18 symbols and 15 arrows. This is a good example of the complex way that symbols are used in many science books. You must be able

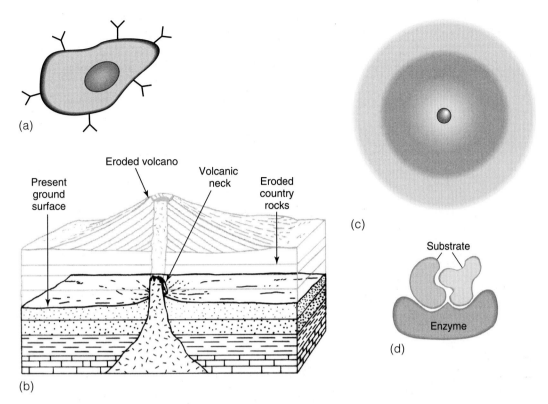

(a)

Present
ground
surface

Eroded volcano

Volcanic
neck

Eroded
country
rocks

(b)

(c)

Substrate

Enzyme

(d)

Figure 9.3 Examples of symbols that must be interpreted. Note how lines, shapes, and shades combine to form individual but often complex symbols. ((b): From Carla W. Montgomery, *Physical Geology,* 3rd edition. Copyright © 1993 McGraw-Hill Company, Inc., Dubuque, Iowa. All Rights Reserved. Reprinted by permission. (c): From Kent M. Van De Graff and Stuart Ira Fox, *Concepts of Human Anatomy and Physiology,* 4th edition. Copyright © 1995 McGraw-Hill Company, Inc., Dubuque, Iowa. All Rights Reserved. Reprinted by permission.)

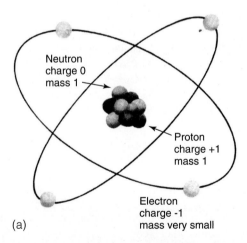

Neutron
charge 0
mass 1

Proton
charge +1
mass 1

Electron
charge -1
mass very small

(a)

Figure 9.4 Arrows are important symbols. Notice in (a), (b), and (c) how arrows are used in different ways to convey different meanings. ((a): From Carla W. Montgomery, *Physical Geology,* 3rd edition. Copyright © 1993 McGraw-Hill Company, Inc., Dubuque, Iowa. All Rights Reserved. Reprinted by permission. (b): From Charles C. Plummer and David McGeary, *Physical Geology,* 7th edition. Copyright © 1996 McGraw-Hill Company, Inc., Dubuque, Iowa. All Rights Reserved. Reprinted by permission. (c): From Sylvia S. Mader, *Biology,* 6th edition. Copyright © 1998 McGraw-Hill Company, Inc., Dubuque, Iowa. All Rights Reserved. Reprinted by permission.)

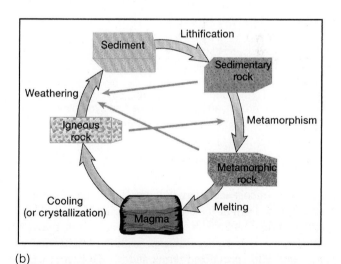

Lithification

Sediment

Sedimentary
rock

Weathering

Metamorphism

Igneous
rock

Metamorphic
rock

Cooling
(or crystallization)

Magma

Melting

(b)

Replication

DNA

Transcription

mRNA

Translation

Protein

(c)

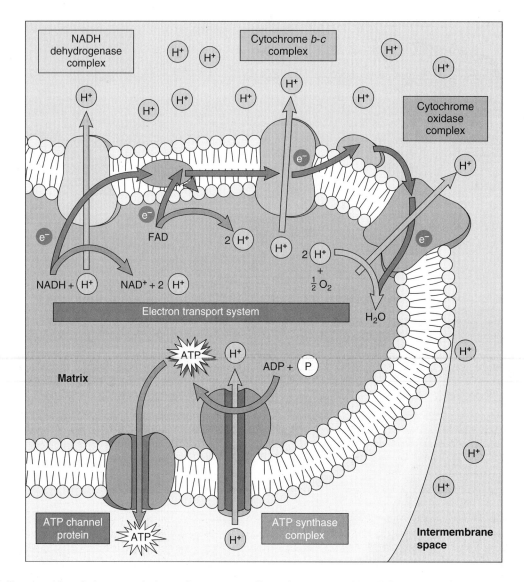

Figure 9.5 Terms, abbreviations, symbols, and arrows are all used to communicate information. This figure demonstrates the use of standardized symbols as well as symbols that must be interpreted. (From Sylvia S. Mader, *Biology,* 6th edition. Copyright © 1998 McGraw-Hill Company Inc., Dubuque, Iowa. All Rights Reserved. Reprinted by permission.)

to find the symbols and interpret them. Remember, a right-brained person might have difficulty seeing the detailed symbols in figure 9.5. What you should realize is that your instructor will communicate science with the English language and with specialized terms and symbols. It is expected that you will learn the terms and symbols of science.

Types of Figures

The types of figures used include maps, illustrations, photographs, diagrams, charts, and graphs. Tables and some symbolized information in texts are not numbered as figures but should be treated as if they were figures.

Maps

Maps are used to relate scientific information to continents, nations, or some other geographic area. Weather, biological, ecological, or geological features are represented on maps (figure 9.6).

Illustrations and Photographs

Illustrations are artistic renditions of something the author thinks will enrich his or her description. Photographs are used to provide views of things described in the text or to perk up your interest. Illustrations and photographs might be placed next to each other for easy comparison (figure 9.7).

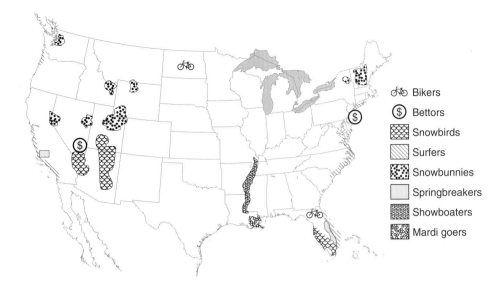

◎♂ Bikers
$ Bettors
▨ Snowbirds
▨ Surfers
▨ Snowbunnies
▨ Springbreakers
▨ Showboaters
▨ Mardi goers

Figure 9.6 Rest and recreation grounds. Humans migrate to various locations around the country for rest and recreation. This map locates approximate R & R areas for migrating humans.

A volcanic neck or pipe—all that remains of a long-eroded volcano. (a) Schematic diagram: Note cylindrical shape and discordant character. (b) An example exposed at the surface by erosion: Devil's Tower, Montana.

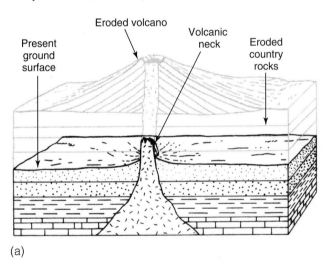

(a)

(b)

Figure 9.7 This is an example of a photograph and diagram. In (a), note that the caption summarizes the information in the body, but does not interpret all the symbols. (Figure from Carla W. Montgomery, *Physical Geology,* 3rd edition. Copyright © 1993 McGraw-Hill Company, Inc., Dubuque, Iowa. All Rights Reserved. Reprinted by permission. Photo: Carla Montgomery.)

Diagrams

Diagrams are simplified illustrations. Components are seldom to scale. Diagrams could be realistic outlines of objects, or they could be generalized impressions of objects. Atoms, electric circuits, telescopes, microscopes, cell parts, organ systems, weather systems, and strata of rock are all examples of things that might be diagrammed. Frequently diagrams express relationships between things. Examples include flow, organization, process, tree, and comparison diagrams. A life cycle, rock cycle (see figure 9.4*b*), or nutrient cycle might be represented in a diagram. Arrows, different-shaped symbols, terms, and simple illustrations might all be included in a diagram (see figure 10.1).

Tables

Tables organize information into rows and columns. Tables are used to compare data or to summarize characteristics of lists of different components. Examples can be found in this chapter (tables 9.1, 9.2, and 9.3).

"Symbolized" Information

The representations of chemical reactions are good examples of these "figures." Atoms and molecules are symbolized in some way along with an arrow to represent the direction of a chemical change. Examples can be found in figures 8.1, 9.4, and 9.5.

Graphs

Graphs usually compare two factors in a qualitative or quantitative fashion. A quick analysis of a graph gives you very specific information or enables you to see relationships between two or more factors. Forms of graphs include bar, pie, line, and pictographs. Each type of graph has its own format. Graphs are titled, units are labelled, and a legend distinguishes information within the graph (see figure 10.3).

Computers and Interactive Software

Computer resources such as CD-ROMs, videodiscs, and course and textbook websites assume you will identify and understand the terms and symbols used in these learning tools. These learning tools must be viewed a number of times to see it all. In these learning tools the symbols frequently move, and you must apply terms to this movement. Some students find these resources helpful. Still other students can't process all that information, and it's

one moving blur. These new techniques of communication add a new dimension and help students envision scientific information. The question is how will they help you learn. You must learn to recognize the terms and symbols used in these tools. Figures on the computer screen should be treated with the same attention that figures in the textbook or lecture are treated.

Review

1. Each science has developed a specialized vocabulary.
2. It is important to learn the terminology and the symbols of science.
3. It is important that you be able to use the vocabulary of science to express complete and logical thoughts.
4. You can use your notes, word inventories, end-of-chapter key terms, and instructor's vocabulary list to identify the terms you need to learn.
5. Vocabulary-definition lists or flash cards help you learn the important terms and symbols.
6. Symbols are used to express the content and concepts of science in shortened form.
7. Symbols take many different forms and have specific meanings. It is important to identify and learn what the symbols mean.
8. Arrows are of particular importance, and you should be able to interpret their meaning.
9. Computer resources use terms and symbols to explain information. Treat these as you would a textbook.

EXERCISE 9

Name _____

Date _____

Terms, Symbols, and Figures

....

1. What is the difference between knowing the meaning of a term and knowing how to use the term?

_____

_____

_____

2. Explain two ways suggested in this guide to learn new vocabulary terms and symbols.

_____

_____

_____

3. How can you identify which terms and symbols you must learn?

_____

_____

4. How can the index and glossary help you learn new terms?

_____

_____

5. Why are symbols used in science?

_____

_____

6. How can a study group help you learn and comprehend new terminology?

_____

_____

7. What units of measurement do the different symbols (different-sized lines and numbers) represent on these rulers?

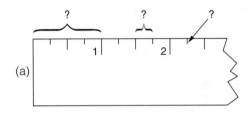

(a)

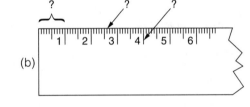

(b)

_____

_____

_____

_____

_____

_____

8. What does each of the 13 symbols represent in this chemical reaction?

$$H_2O + CO_2 \rightarrow H_2CO_3$$

_____

_____

_____

_____

9. **Challenge yourself:** Answer these questions that refer to the following figure. You might want to refer to information in figure 8.1.

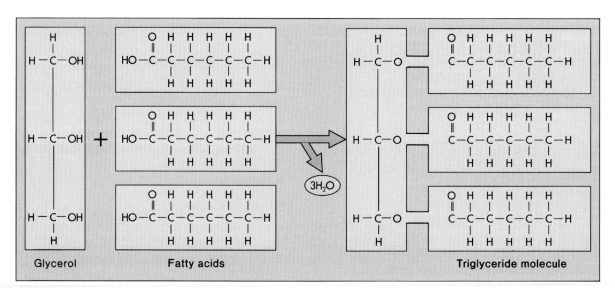

Structure of a fat molecule. This fat molecule, a triglyceride, is formed by dehydration synthesis in which the glycerol is attached to three fatty acids. (From Peter H. Raven and George B. Johnson, *Understanding Biology,* 3rd edition. Copyright © 1995 McGraw-Hill Company, Inc., Dubuque, Iowa. All Rights Reserved. Reprinted by permission.)

9. a. What do the various letters represent?

_____

_____

_____

9. b. What does the large "+" represent?

_____

9. c. Why did the artist create boxes around the molecules?

_____

_____

9. d. What do the "3" and the "2" represent?

_____

_____

9. e. Where specifically do the atoms in the water molecules come from?

_____

_____

9. f. How many fatty acid molecules are symbolized?

_____

9. g. How many atoms are in glycerol or fatty acid molecules?

_____

9. h. What process does the branched arrow represent?

_____

9. i. How many molecular parts went into the synthesis of a triglyceride molecule?

_____

9. j. What do the dashed lines between the letters represent?

_____

_____

9. k. What does the "tri" and "glycer" refer to in the term "triglyceride"?

_____

10. **Challenge yourself:** Answer the questions about the arrows in this figure.

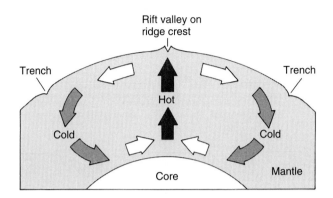

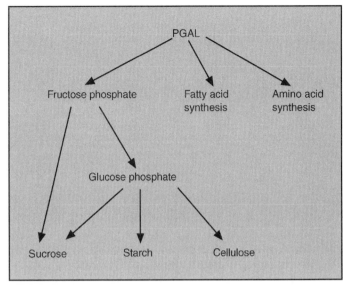

Sea-floor spreading. (a) Hess proposed that convection extended throughout the mantle. (Scale of ridge and trenches is exaggerated.) (b) Hot mantle rock rising beneath the mid-oceanic ridge (a spreading center) causes basaltic volcanism and high heat flow. Divergence of sea floor splits open the rift valley and causes shallow-focus earthquakes (stars on ridge). Sinking of cold rock causes subduction of older sea floor at trenches, producing Benioff zones of earthquakes and andesitic magma. (Figure and legend from Charles C. Plummer and David McGeary, *Physical Geology,* 7th edition. Copyright © 1996 McGraw-Hill Company, Inc., Dubuque, Iowa. All Rights Reserved. Reprinted by permission.)

10. a. What do the 10 arrows represent?

_____

_____

10. b. Why are the arrows different shades?

_____

_____

10. c. How do the key terms in the caption help interpret the symbols and arrows in the body of the figure?

_____

_____

_____

_____

11. **Challenge yourself:** Different terms and arrows are included in the following figure. Answer these questions.

11. a. What do the three arrows pointing away from PGAL (phosphoglyceraldehyde) mean? (PGAL is a molecule.)

_____

_____

11. b. What do the other five arrows represent?

_____

_____

_____

_____

11. c. Where do "P," "G," and "AL" in PGAL come from in that "monster" scientific term?

_____

_____

11. d. Which of the molecules in this figure would be considered monomers or polymers? (Refer to figure 8.1.)

_____

_____

_____

11. e. Create symbolic shapes to represent each of the eight molecules in this figure. Anything you draw will be correct, but they should be different sizes and shapes and should reflect the information in the figure. You might also want to refer to figure 8.1. Label the symbols you have drawn.

Analyzing Figures

Objectives

When you have read this chapter, you will be able to answer these questions:

1. How can I analyze figures so they make sense?
2. Can figures help me visualize the concepts?
3. How can I create my own figures, concept maps, graphs, and tables to help me learn all that "stuff"?
4. Are figures "worth a thousand words"?

F lip through any science book and you will see that **figures** are used to communicate scientific information. Figures will help you visualize the instructor's spoken word and the textbook's written word. It is important that you have the study skills to comprehend figures. In other words, it is important to be "figure-literate." Instructors assume that "a picture is worth a thousand words" and that you can comprehend figures in textbooks. In addition, they expect you to understand the figures they create even if the figures are poorly drawn.

Make every effort to understand the figures used by your instructor. All too often, students ignore or skip over these valuable study aids because they look too "difficult." The most important figures are those used in the lecture and lab. Part of the survey and reading of S + Q + 3R + P should be the analysis of figures in your textbook or notes. When you recite and review, you should be able to explain or create figures. Part of the "P" during study sessions should be deciding the relationships between the content of important figures and the content of class notes. Doing all this takes time . . . that's why instructors expect you to spend 6 to 12 hours a week studying science.

Remember, if your instructor uses figures from the textbook, then bring your book to class. Follow the discussion of the figure in your textbook. As the instructor mentions the information in the figure, place a check mark next to it. If you do this, you will be a better listener and will not get bogged

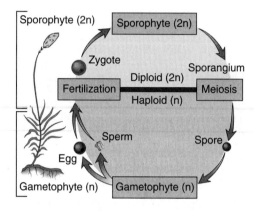

Figure 10.1 The body of this figure uses illustrations, terms, abbreviations, and arrows. Note that the body has direction and isolates four terms in boxes. (Figure from Sylvia S. Mader, *Biology,* 6th edition. Copyright © 1998 McGraw-Hill Company, Inc., Dubuque, Iowa. All Rights Reserved. Reprinted by permission.)

down in trying to draw detailed figures. Your attention will be focused on learning information rather than drawing!

Figure Analysis

Each figure has a caption and a main body. The caption gives the number of the figure and briefly describes the figure. The body depicts the content as explained in the text and caption. The body contains different kinds of symbols. Assume all lines, colors, shapes, arrows, and marks have meaning. More information is added when letters, numbers, or terms are used (figure 10.1). Some people have trouble "reading" figures because they don't see all of the symbols. Some figures have 30 to 50 symbols (see figure 9.5). Part of figure literacy is to pay attention to the details. In addition, close examination of the symbols used in many figures often yields more information than is included in the caption.

When you analyze a figure, have a pencil in hand to help you "get into" the details of the figure. Use the S + Q + 3R + P technique. Follow these six steps:

I. **Survey**
   A. Caption
      1. Underline and define the key words.
   B. Body
      1. Scan to determine direction and identify types of symbols.
      2. What is this figure about?
      3. How many "bits" of information are in the figure? (Yes, point out and count the shapes, colors, labels, arrows, etc. This counting forces you to examine the details.)

II. **Question**
   A. Caption
      1. Does the caption point out specific elements in the figure?
   B. Body
      1. What elements stand out in the body of the figure?

III. **Read**
   A. Caption
      1. Examine carefully, for detail.
   B. Body
      1. Translate or interpret the symbols into terms.
      2. Connect the symbols to each other.
      3. Relate the caption to the body. Match the terms in the caption to the symbols in the body.
      4. Use your pencil to point out all other parts of the figure not identified. Try to decide what they represent.

IV. **Recite**
   A. Create an essay of a "thousand" words by interconnecting the terms with linking words.
   B. Recite the content of the figure once again.
   C. Answer questions you asked.
   D. Relate the figure's content to the material covered in your notes or in the laboratory manual.

V. **Review**
   A. Review the content of both the caption and body of the figure.

VI. **Practice**
   A. Explain the figure to yourself.
   B. Reproduce the figure on a blank sheet of paper.

Create a "Figure Analysis Study Card" by listing on a flash card the six steps outlined above. Refer to it as you analyze figures. Place it in your textbook as a handy guide to develop your "figure literacy" skill.

Computer Graphics

The graphics-enriching computer software contains the same elements as figures in a textbook. Some of the graphics are static (not moving); others have symbols that move. Rather than using a pencil, you use a cursor to interact with the figure. The dynamic nature of interactive software automatically involves the learner in the graphics. A press of a button will lead to more detailed information, related information, or questions about the graphics. The successful use of computer graphics depends on your ability to recognize the symbols and terms. You must be willing to explore the software pathways but must identify the parts that will help you learn the content of the course. You also need a system to understand the graphics. Thus the S + Q + 3R + P technique should be used as you work with computer graphics.

Creating Figures

Developing the ability to represent written information in a figure format is another important study skill. If you create a figure representing something, you will be reworking the information into your own thinking. Creating figures will give you practice using information and will be a way to bridge the top and bottom part of the learning pyramid (see figure 3.1). The symbols you choose to use are defined by you. Any shape you draw can represent a molecule or a cell just as long as you label it and understand what you are representing (see Exercise 9, question 11e.). You can draw arrows or lines; all you have to do is link an appropriate term with the line or arrow. The more you practice this skill, the easier it gets. An example helps you understand the value of this study skill.

A heading and paragraph in Sylvia Mader's *Biology*[1] reads:

Monoclonal Antibodies Have Same Specificity
As previously discussed, every plasma cell derived from the same B cell secretes antibodies against a specific antigen. These are monoclonal antibodies (Gk. mono, one; clon, branch; and anti, against) because all of them are the same type and because they are produced by plasma cells derived from the same B cell.

---

[1] Sylvia S. Mader, *Biology,* 5th edition, p. 663, 1996. Times Mirror Higher Education Group, Inc., Dubuque, IA.

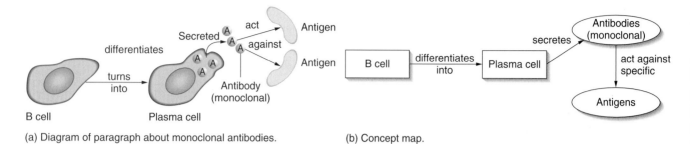

(a) Diagram of paragraph about monoclonal antibodies.   (b) Concept map.

Figure 10.2 The same information can be symbolized in either (a) a diagram or (b) a simple concept map.

Figure 10.2a gives an example of the kind of figure you might create from this heading and paragraph. Note the use of various symbols, arrows, and words to create a figure. Do you think the figure is a fair representation of the paragraph?

Concept Maps

A concept map, also called a flowchart, is a visual display that depicts relationships among concepts and attempts to simplify densely written scientific sentences (figure 10.2b). Note that figure 10.2 depicts the same information in two ways. Part (a) is a diagram using symbols to represent cells and molecules. Labelled arrows give direction and identify structures. Part (b) symbolizes the same information, isolates concepts, and connects these concepts with lines and linking terms, but it does not attempt to symbolize molecules or cells.

Concepts are isolated in boxes or circles, and lines are drawn and labelled to interconnect related information or concepts. The encircled concepts can be placed anywhere on the paper. If you use this study skill, then you will develop a "feel" of where to place things. It is the connecting lines with correctly labelled linking terms that are most challenging. The map develops a weblike appearance as more concepts are interconnected. Sometimes, particularly at first, the maps have to be redrawn because the orientation on the page is not quite right. (Just like when you draw a "road map" with directions to get to your house.)

To some, this mapping seems chaotic and confusing. To those who develop the skill and create the maps, it provides a way to review and summarize information about a topic. In addition, concept mapping will allow you to think in multiple directions and relate abstract and concrete thoughts.

The freedom of creation allowed by concept maps enables you to organize the information in your own way. Some science instructors might not be comfortable or familiar with this skill. Since concept maps are not linear or sequential, it does not appeal to a left-brained person, but it probably does work for a right-brained person. Developing this skill will help you learn because it provides another way to connect the top and bottom of the learning pyramid (see chapter 3).

Remember, "a picture is worth a thousand words" only if you are figure-literate.

Graphs

Analyze graphs by using the above figure analysis technique. Students tend to have little experience or practice in constructing graphs. If graphs are required, find out what format or style the instructor expects. The most common kinds of graphs are line (figure 10.3), bar (figure 6.1), and pie (figure 2.1). All graphs must include the following: title, legend, and axis and unit labels. If you are uncertain as to the proper format, seek help from your tutor or find examples of the type of graph in a textbook and use it as a model. Keep the following in mind when you construct a line graph (figure 10.3):

1. Use the whole sheet of graph paper, not just a corner. This means that units must be scaled out to the ends of each axis.
2. Create a "title box" that includes a graph title and your name.
3. Draw each axis two to three centimeters from the bottom and left side of the edge of the graph. The y-axis is the vertical axis, and the x-axis is the horizontal axis.
4. The origin is the point where the x- and y-axes meet.
5. Divide each axis at regular intervals using as much of the axis as possible. Number the division marks clearly.
6. Label each axis with units and the type of thing represented by the unit. These should be clearly written and centered along the axes.
7. Mark the data points with simple but visible dots. Some instructors request that data points be surrounded by circles, triangles, or squares if more than one data line appears on the graph. If you use more than one symbol, label the meaning of these symbols in a legend. If you misplace the dots, try to erase them, but if the paper is marred, start over.
8. Some instructors allow "play connect the dots"; others say "never connect the dots." Connect dots if the instructor indicates it. On the other hand, draw the line or curve that best "fits" all

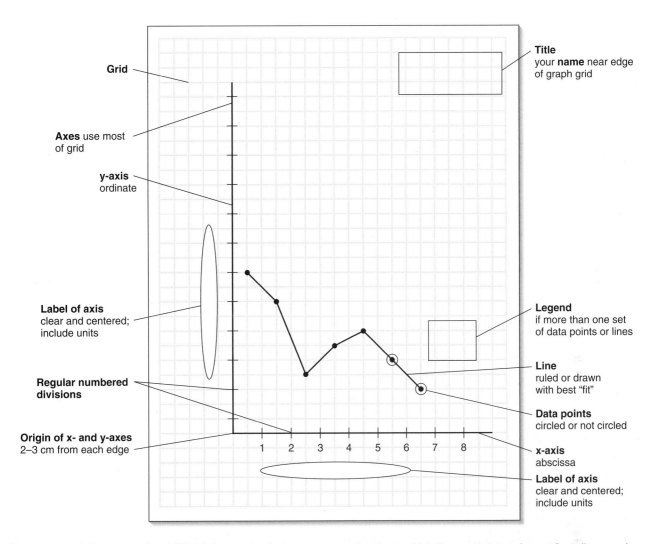

Figure 10.3 A line graph should be well-organized, clear, neat, and accurate. This figure depicts a format for a line graph.

the data points if that is what is expected. Use rulers or French curves.

9. Lines should generally be drawn only from one data point to another. However, some styles of line graphs extend the line slightly beyond the first and last data points.

10. Work neatly.

11. If you have trouble with the technique of making a graph, seek help from the instructor or science tutor.

The most common graphing errors, with the accompanying loss of grade points, are:

* omission of the title,
* improper axis title,
* improper representation of units,
* confusing data labels, and
* omission of legend.

These are some of the most common graphing errors. As previously discussed, instructors usually have a specific graph format and will mark your work according to the expected format. *Be sure to ask for your course's expected format for graphs and follow it.*

Computer Creation of Graphs Graphs can be constructed using spreadsheet programs. The software prompts you through various options or will construct a graph automatically from the data put into the spreadsheet. For many graphing programs, you must enter the data correctly or the graph will be inaccurate. If you want to use the computer for graphing but do not know how to input data, seek help from your instructor, an on-campus learning center, or a tutor. Caution must be taken to insure that the data is graphed correctly and that all of the elements indicated in figure 10.3 and specified by your instructor are included in your computer-created graph.

Tables

Remember Exercise 8 asked you to S + Q + 3R + P the information in figure 8.1. A table was also part of that figure.

Chemistry — an information table:

Gas Laws:

Name	relationship	symbols	description	application
Boyle's	$PV = k$ (constant T)	P = pressure V = volume k = constant i = initial f = final	At constant temp., the pressure and vol. of a sample of gas are <u>inversely</u> proportional.	If conditions on sample change then $P_i V_i = P_f V_f$
Charles	$V = kT$ (constant press.)	V = volume T = temp. (°K) k = constant i = initial f = final	At a constant pressure the volume of a gas is <u>directly</u> proportional to its absolute temp.	If either the V or T conditions change then: $\dfrac{V_i}{T_i} = \dfrac{V_f}{V_f}$

Figure 10.4 Creating your own information table is a valuable way to summarize the content of your course.

Note the title, labels of columns, terms naming the rows, and examples of the terms mentioned in the rows. The table also relates the type of macromolecule with its subunit and function. There are 40 or so "bits" of information in this one table. How much of this would you be expected to know? The answer is "If your instructor mentions it, then you must learn it."

As you try to summarize the things you must learn, it is often helpful to construct your own information tables. If you do this, you will be reorganizing information in your own way. You will be practicing using the information! The difficulty in organizing tables is deciding what to name the columns and rows. Once this is done, all that remains is to fill in the information from your notes, textbook, or laboratory work. Examine samples of tables in the textbook to help organize your own tables. Figure 10.4 displays an information table created for topics in chemistry.

Review

1. Figures are important aids to communicate scientific information.
2. You should develop and practice the skills that will enable you to analyze and comprehend different types of figures.
3. Figures include a number, a caption, and a body of information.
4. When you analyze a figure you should:
   - survey the figure;
   - count the "bits" of information;
   - identify the key terms in the caption;
   - create questions;
   - read the caption;
   - translate symbols into terms;
   - relate the caption to the figure;
   - recite how the figure is worth a thousand words;
   - identify the unlabelled parts of the figure;
   - relate the figure to the lecture or laboratory content;
   - answer the questions; and
   - practice by redrawing or explaining the figure.
5. The ability to create figures, concept maps, graphs, and tables is an important study skill.

EXERCISE 10

Analyzing Figures

....

1. Check which of the following you do:

_____ Correct or redraw figures drawn in class notes.

_____ Compare figures in notes to figures in the text-book or laboratory manual.

_____ Decide which figures in the text to analyze and learn.

_____ Explain figures in your notes or textbook to someone.

_____ Survey the caption and body of a figure.

_____ Count the "bits" of information in a figure.

_____ Create questions about the terms, symbols, or information in the caption or body.

_____ Recall or define the key terms in the caption or body.

_____ Create "a thousand words" after analyzing the figure.

_____ Answer the questions you asked about the figure.

_____ Relate the information in a figure to the contents of your notes.

_____ Examine figures as they are referred to when reading the textbook.

_____ Create symbols and figures.

_____ Create tables.

_____ Create concept maps.

_____ Compose graphs with the proper format.

2. What actions will you take to improve your "figure literacy"?

_____

_____

_____

3. Test your memory. List six steps and activities suggested in this chapter to analyze figures.

_____

_____

_____

_____

_____

4. Analyze the descriptive figure below. The following questions will help you develop the "system of figure analysis." Record your answers on a sheet of paper. Refer to Appendix B for more information about this question.

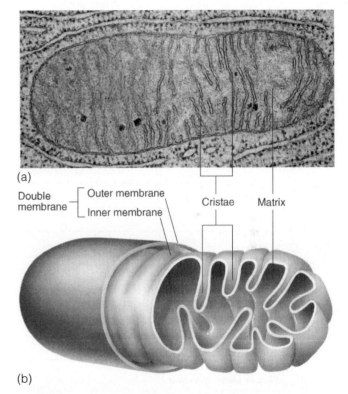

Mitochondrion structure. (a) Electron micrograph. Magnification, × 70,000. (b) Generalized drawing in which the outer membrane and portions of the inner membrane have been cut away to reveal the cristae. (Figure and legend from Sylvia S. Mader, *Biology,* 5th edition. Copyright © 1996 McGraw-Hill Company, Inc., Dubuque, Iowa. All Rights Reserved. Reprinted by permission. Photo: Courtesy of Dr. Keith Porter.)

4. a. What is this figure about?

_____

_____

_____

4. b.   How many "bits" of information are in the figure?

_____

4. c.   What do you consider to be the key terms in the caption?

_____

_____

_____

_____

4. d.   What kind of photograph is in the figure?

_____

4. e.   How many membranes does this structure have?

_____

4. f.   The inner membranes of this structure are folded inward to form structures called

_____.

4. g.   The inside space of this structure is called the

_____.

4. h.   Describe the structure of a mitochondrion in a "thousand words or less."

_____

_____

_____

_____

_____

5.   Analyze figure 9.7. Answer the following questions:

5. a.   Where is Devil's Tower?

_____

5. b.   What is Devil's Tower?

_____

_____

5. c.   Explain how Devil's Tower was formed.

_____

_____

_____

_____

_____

_____

_____

6.   Analyze the following map and bar graphs. Answer questions a–f.

6. a.   What states were involved in the Great Flood of 1993?

_____

6. b.   What do the numbers on the boundaries of the map represent? (The author assumes you will know this.)

_____

6. c.   What do these bar graphs relate?

_____

_____

_____

6. d.   What rate is indicated on the y-axis (ordinate or vertical line) of one of the graphs?

_____

6. e.   What unit of measurement is indicated on the x-axis (abscissa or horizontal line) in each graph?

_____

6. f.   The caption serves as titles for the graphs. Do you detect any errors in the format of these graphs?

_____

_____

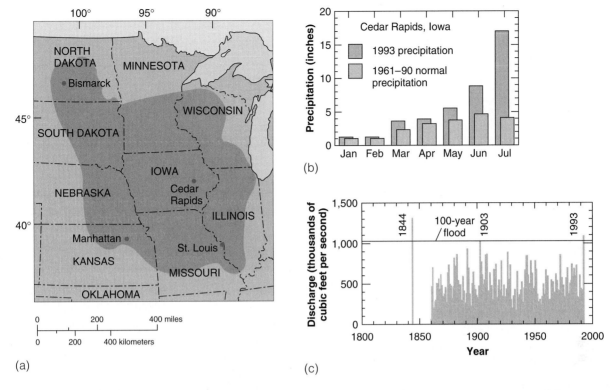

The Great Flood of 1993. (a) Area of flood. (b) 1993 rainfall at Cedar Rapids, Iowa, compared to normal rainfall.
(c) Discharge of the Mississipi River at St. Louis compared to 100-year flood. (Figure and legend from Charles C. Plummer and
David McGeary, *Physical Geology,* 7th edition. Copyright © 1996 McGraw-Hill Company, Inc., Dubuque, Iowa. All Rights Reserved. Reprinted by permission.

7. **Challenge yourself:** Create a diagram for the following statement. Use whatever symbols you like.

Amino acids are small molecules with a simple basic structure: They contain an amino group (—NH$_2$), a carboxyl group (—COOH), a hydrogen atom (H), and a functional, or "side group," designated R, all bonded to a central carbon atom (C).[2]

8. **Challenge yourself:** Create a simple diagram and concept map for the following statement:

Phagocytosis (from the Greek *phagein,* meaning "to eat," and *cytos,* meaning "cell,") is when material brought into the cell is an organism or some other fragment of organic matter.[3]

[2] Peter H. Raven and George B. Johnson, *Understanding Biology,* 3rd edition, p. 40, © 1995 Times Mirror Higher Education Group, Inc., Dubuque, IA.

[3] Raven and Johnson, pp. 92 and 93.

CHAPTER 11

Practice Understanding Figures

Objectives

When you have read this chapter, you will be able to answer these questions:

1. Just how good am I at figuring out scientific figures?
2. Can I create figures, symbols, and graphs?

This chapter will allow you to practice the skills discussed in chapter 9 and chapter 10. The ability to understand and create figures will help you comprehend science. As you practice figuring out figures, remember that "the devil is in the details."

Answering the Questions

This chapter consists primarily of exercises designed to give you practice in analyzing figures. Some clues to answers can be found in Appendix B. The answers that are given try to demonstrate "good technique" for answering questions. Compare the way you answer the questions with the style of answers given. When answering questions, some students take the easy way out and answer all questions with one word. Sometimes this is appropriate; other times a more complete answer is required. That is, a complete sentence or short essay might be necessary. Writing complete thoughts is a chance to practice linking terms in meaningful ways. In addition, by creating complete thoughts in sentences and paragraphs, you can demonstrate your understanding of the information. (See chapter 12 for more complete coverage of this topic.) Answers should:

- be given in full sentences;
- use the words and terms of the question; and
- apply the terms in the figure.

"Figuring Out" Figures

As you progress through these exercises, develop the skill to identify key terms, see and interpret symbols, "create a thousand words," and understand the diagrams. The exercises vary in their approach. Some give you a figure and questions to guide your analysis. Others ask you to create figures, questions, or graphs. Still others ask you to "create a thousand words." The intent of these exercises is to improve skills that will enable you to **learn how to learn** science and to become **figure-literate.** Of course, as you do this, you'll learn a bit of science along the way.

Use your "Figure Analysis Study Card" (chapter 10) to guide your study. As you analyze the figures, remember to survey, scan, underline the key terms, create questions, etc. The learning objective is to develop a system of study to improve your figure literacy.

Review

1. Practicing a study skill will help students understand information and figure out the details of what they must learn.
2. Figure literacy will help you comprehend science.
3. There is a proper format to follow when answering questions, as exemplified in the answers given in Appendix B.

EXERCISE 11a

Figuring Out a Photograph

....

Refer to the photograph to complete this exercise.

1. What is this photo about?

   _____

   _____

   _____

2. How many bits of information are in the photo?
   (Remember a "bit" is any symbol, arrow, or label in
   the body as well as the terms in the caption.)

   _____

3. What are the key terms in the caption?

   _____

   _____

   _____

   _____

   _____

4. Define "seed," "germination," "shoot," and "root."

   _____

   _____

   _____

   _____

   _____

   _____

5. What parts grow out of a seed?

   _____

   _____

Upon germination of a wheat seed, growth and
development result in a shoot growing upward and
roots growing downward. (© Adam Hart-Davis/SPL/Photo
Researchers, Inc.)

6. Which part of the seed is longer in this photograph?

   _____

7. Summarize the contents of this photo in a short essay.

   _____

   _____

   _____

   _____

   _____

   _____

   _____

EXERCISE 11b

Chemical Symbols and Reactions

....

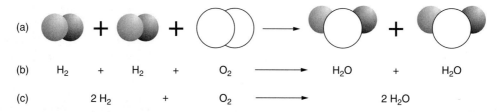

(a)

(b) H_2 + H_2 + O_2 $\longrightarrow$ H_2O + H_2O

(c) $2\,H_2$ + O_2 $\longrightarrow$ $2\,H_2O$

The meaning of a balanced chemical equation is when the number of atoms (by themselves or in molecules) that appear to the left is equal to the number of atoms on the right of a chemical equation. The atoms are rearranged in a chemical reaction.

Before you answer the questions below, use the figure shown here to spend time practicing the steps of analyzing a figure. Make up and record your own questions on a separate sheet of paper. Answer all of the questions. Then compare your questions to those provided here.

1. What is this figure about?

2. How many bits of information are there?

3. Did you underline the key terms in the caption? What are the key terms?

4. How do the questions you created compare with those on this page?

5. What is a balanced equation?

6. What symbol separates the equation into left-hand and right-hand sides?

7. In a balanced equation, what is the relationship of the numbers of atoms on the left-hand and the right-hand sides?

8. What do the symbols H, H_2, O, O_2, and H_2O represent?

9. How does the illustrator represent H_2, O_2, and H_2O?

10. What do the (+) and the → represent?

11. Why are there three representations (a, b, and c) in this figure?

12. What does the 2 in front of H_2 and H_2O represent in part c of figure 11.7?

13. How are the atoms rearranged in this chemical reaction?

14. What are the key terms that might appear on a test?

15. Summarize which numbers and types of reactants will form with new numbers and types of products in this balanced chemical reaction.

16. Relate the cartoon to the scientific figure.

EXERCISE 11C

Name _____

Date _____

Diagrams, Symbols, and Terms

....

The following simple diagrams, chemical symbols, and words depict information about the formation of a polymer. Other information about this representation can be reviewed by referring to figure 8.1.

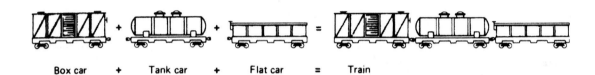

1. Compose a caption for the body of information given in this figure.

2. On a separate piece of paper, make up five questions about both the caption and body of the figure. Answer your own questions.

3. Why are different railroad cars shown in this figure?

    _____

    _____

4. What does the symbol "=" mean?

    _____

5. What do the two arrows indicate has happened in this figure?

    _____

6. What is the name of the process indicated by the two arrows? (Refer to figure 8.1.)

    _____

7. A train is made of railroad cars; a protein is made of a string of

    _____.

8. Where do the hydrogen and oxygen atoms come from to form the two water molecules?

    _____

    _____

EXERCISE 11d

Name _____

Date _____

Tables

....

Most tables contain information that summarizes a topic. This exercise gives you practice finding information in a table.

Table 38.2 Hormones of Digestion

Hormone	Source	Stimulus	Action	Note
Gastrin	Pyloric portion of stomach	Entry of food into stomach	Secretion of HCl	Unusual in that it acts on same organ that secretes it
Cholecystokinin (CCK)	Duodenum	Arrival of food in small intestine	Stimulates gallbladder contraction, and so the release of bile into intestine; stimulates secretion of digestive enzymes by pancreas	CCK bears a striking structural resemblance to gastrin
Secretin	Duodenum	HCl in duodenum	Stimulates pancreas to secrete bicarbonate, which neutralizes stomach acid	The first hormone to be discovered (1902)

Questions   **Answers**

1. How many columns are in this table?

2. How many rows are in this table?

3. What are the five titles of the columns?

4. Where is gastrin produced?

5. What is the action of gastrin?

6. What does CCK mean, and what is CCK?

7. What and when was the first hormone discovered?

8. What stimulates secretin to be secreted?

9. Review the table in figure 8.1 and make up five questions about the information in the table. Record these on the left-hand side of a piece of paper. Answer the questions on the right-hand side next to the question.

E X E R C I S E 11e

Name _____

Date _____

Line Graphs

Remember that line graphs use lines to show how quantities change with a change in time, distance, or some other factor. Use the line graph shown here to complete the questions.

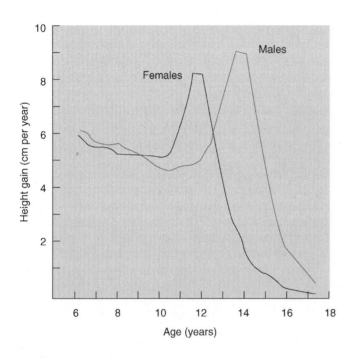

Growth in height of females and males as a function of age. Notice that the growth spurt during puberty occurs at an earlier age in females than in males. (Figure and legend from Kent M. Van De Graaff and Stuart Ira Fox, *Concepts of Human Anatomy and Physiology,* 4th edition. Copyright © 1995 McGraw-Hill Company, Inc., Dubuque, Iowa. © The Society for Research in Child Development, Inc. Reprinted by permission.)

Questions

1. What are the units of time?

2. What does the x-axis represent?

3. What does each increment on the y-axis represent?

4. What is the expected rate of height gain for a 12-year-old female?

5. What is the number of years separating the maximum rate of height gain for females and males?

6. What title would you give to this graph?

7. How does the format of this graph compare to the directions given in chapter 10 (see figure 10.3)?

Answers

EXERCISE 11f

Constructing a Graph

....

The following are data from a laboratory exercise that determined the absorbance (proportion of light hitting a sample that is absorbed) of various wavelengths of light (nanometers) by a combination of pigment molecules extracted from leaves of an *Elodea* plant. Results from replications of experiments are averaged. This average is used to construct a graph of the results. Additionally, calculations are made to represent the variation surrounding the average (such as standard deviation). These ranges are also graphed.

Construct a line graph that plots the absorbance on the y-axis and the wavelengths of light on the x-axis. Follow the format given on page 70 and figure 10.3. For this exercise you can choose to plot:

1. a single replicate;

2. the average of the three replicates; or

3. the average and the surrounding variation (for example, use standard deviation).

Wavelength (nanometers)	Absorbance Replicates		
	(1)	(2)	(3)
350	0.40	0.38	0.44
375	0.41	0.41	0.40
400	0.57	0.64	0.60
425	0.59	0.58	0.60
450	0.44	0.42	0.43
475	0.28	0.26	0.33
500	0.09	0.05	0.09
525	0.05	0.06	0.04
550	0.04	0.05	0.04
575	0.05	0.05	0.05
600	0.06	0.09	0.04
625	0.08	0.08	0.04
650	0.15	0.14	0.16
675	0.28	0.27	0.27
700	0.17	0.11	0.22
725	0.33	0.33	0.35
750	0.38	0.39	0.41
775	0.40	0.41	0.39
800	0.44	0.45	0.46

EXERCISE 11g

Creating Figures

....

1. Create a figure with a caption and body that symbolize the atom described in the following paragraphs.

   All matter is composed of small particles called atoms. The accepted view of an atom is that they are made of a small but dense nucleus and "clouds" of orbiting subatomic particles. These moving particles are called electrons. Their number varies in each kind of atom, but all electrons have negative (–) charges. The nucleus is made of two different kinds of subatomic particles, protons and neutrons. Protons have positive (+) charges, and their number is generally the same as the electrons. Neutrons are not charged.

   A carbon atom has six protons and six neutrons in its nucleus. There are six electrons. Two electrons orbit the nucleus in a "lower" cloud level; the remaining four electrons rotate in a cloud farther out from the nucleus.

   Compare your creation with that in figure 9.4a.

2. Create a figure with a caption and body that symbolize the molecule described in the following paragraph.

   An enzyme is a type of protein that is globular in shape. The irregular surface generally has one major indentation or pocket that is called an active site. The active site is the place where the chemical reactions catalyzed by the enzyme occur.

   Compare your creation with that in figure 9.3d.

EXERCISE 11h

Name _____

Date _____

Creating an Essay from a Figure

....

Analyze the following figure using the S + Q + 3R + P technique of figure analysis that you have been practicing (see p. 68). After you have "learned" the figure, practice using the information by composing an essay about it. This is a challenging figure, but worth the effort since it will focus you on how to improve your figure analysis skills for your science course.

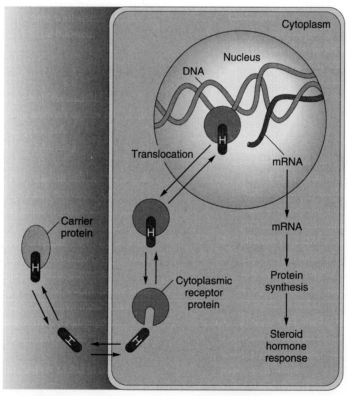

The mechanism of the action of a steroid hormone (H) on the target cells. (Figure and legend from Kent M. Van De Graaff and Stuart Ira Fox, *Concepts of Human Anatomy and Physiology,* 4th edition. Copyright © 1995 McGraw-Hill Company, Inc., Dubuque, Iowa. © The Society for Research in Child Development, Inc. All Rights Reserved. Reprinted by permission.)

EXERCISE 11i

Flowcharts and Problem-Solving

....

The following flowchart describes a pathway of actions to help you solve math-based problems.

1. Read through the set of directions illustrated in the flowchart.

2. Compare the charted technique for solving a problem to the technique that you use to solve a problem. List the similarities and differences.

_____

_____

_____

_____

_____

_____

_____

3. Briefly discuss if the pathway of actions illustrated in the flowchart can help you to solve math-based problems.

_____

_____

_____

_____

4. After you have read through the flowchart, apply the pathway to help you solve the following question.

Problem:

A label on a bag of potatoes indicates that a serving size weighs 148 grams. It also indicates that one serving size contains 3 grams of protein, 27 grams of starch and sugars, and 0 grams of fat. Chemical tests were performed on the cellular extracts of a 5-gram piece of potato. The tests indicated the presence of protein and starch and sugars. How many grams of protein and grams of starch/sugars can be found in the 5 grams of potato?

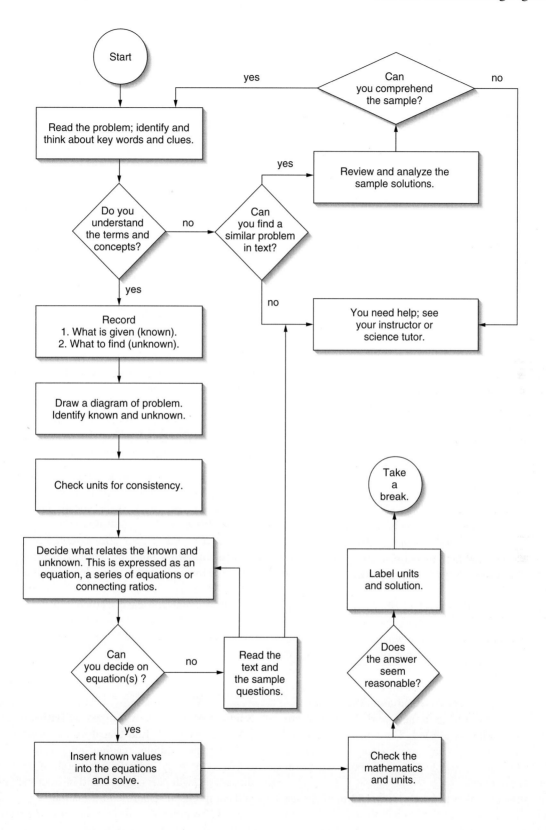

CHAPTER 12

Assignments and Reports

Objectives

When you have read this chapter, you will be able to answer these questions:

1. If the teacher doesn't give any assignments, does that mean I don't have to study?
2. Why read the textbook or lab manual—the instructors will tell us what to study, won't they?
3. What kinds of assignments can I expect?
4. What's expected in a lab report?
5. When assignments are returned, what should I do with them?

In high school, students are given homework to direct their study. Notes that are given are memorized. Tests seem to be taken directly from the homework and notes. If you do the work, you'll get decent grades. Extra-credit projects might be available to increase scores. Since everything is rather concrete, a variety of study skills need not be used or developed.

When students come to college, they expect the same treatment. Surprise! Things are **not** the same at college. Students are not told time and again what to study. Reading assignments are very general and often overwhelming. Laboratories are done; sometimes written work is collected, and other times nothing is checked. In other cases, formal reports and assignments are collected and marked, but students aren't reminded to hand in their work. Test dates are set. Topics and readings are listed. The types of questions to be asked are announced, but a review for the test is not routine. Retesting to improve low scores is rare, and extra-credit projects even rarer.

Some students study to barely pass, and they do just that. Others decide they want an A, and they work to earn one. Still other students think they are studying enough to get "at least a B" but wind up with a D or F and wonder why. These students don't recognize that they are not adjusting to college study. They desire good grades, but don't have the skills to achieve these grades.

The **academic behavior** expected at college has to do with connecting the way in which you learn and the way in which college is taught (see chapter 3). The teachers at college will lead you to what must be learned, but you must decide how much you will learn. The studying, reading, and work you complete are up to you. Each student will develop an individual learning style that is based on his or her goals and motivation.

Assignments

Assignments vary from the general to the specific. Note **studying** and chapter **readings** are examples of general assignments. **Word lists, study sheets, laboratory reports,** and **problem-solving questions** are categories of specific assignments. All of your efforts should be directed toward **identifying** what must be learned, **organizing** a plan to learn the material, and actually **learning** the material. Writing assignments and problem-solving allow you to **apply** the information you have learned. All these activities help prepare you for tests. Problem-solving (essays and calculations) and tests will be discussed more completely in chapters 13 and 14.

Studying

Study is a series of **organized activities** directed toward learning. You will find that if you engage in previewing; listening; note-taking; reading; reviewing; reciting; recalling; group study; questions generation; self-testing; and discussion with peers, tutors, and instructors, you will learn and comprehend scientific information.

Instructors encourage you to and assume you will study. They do not check to see if you are studying. The fact that they don't give a concrete homework assignment to be handed in does not mean that the information is not important. It is assumed you are an adult and will study to learn the information. Remember, your own self-direction and self-discipline are key to studying. Keep in mind the importance of time management and establishing a list of study objectives.

Chapter 3 and chapter 7 describe the study skills and activities you should do as routine, self-directed, and self-disciplined assignments. Review the guidelines of what is important to study in chapter 7.

Reading

Reading assignments are generally included in the course syllabus. This may likely be the only place that these assignments are mentioned, so frequently refer to your course syllabus for your reading assignments. Some instructors will repeat the assigned readings by noting them on the board or by specifying certain parts of the chapter to be read. For instance, they might say "Read chapter 3, but pay particular attention to pages 134–163." That's a study hint; record it in your notes! Laboratory exercises will be assigned. "Outside" readings might well be assigned. It is assumed you will relate your readings to what is being presented in lecture or lab. The instructor's style (see chapter 2) will influence when and how you read your textbook.

The assigned readings will help you learn the information presented in class. The explanations, figures, tables, and summaries included in the text and manuals will reinforce the information you listened to in class. However, the only checks on whether or not you have done the reading to complement and supplement your class work will be tests, laboratory reports, or other written assignments.

Remember to use the reading technique of S + Q + 3R + P explained in chapter 8.

Hein, Best, and Miner state, "Since your laboratory time is limited, it is important to carefully study and clearly understand the experiment scheduled for a laboratory period *before* you come to the lab. Without at least one hour of preparation before each lab—and this should be considered a standing homework assignment—it is doubtful that the experiment and its report form can be satisfactorily completed in the allotted period of time."[1]

Knowing about an error students make might prevent you from making the same **"reading assignment error."** The course syllabus gives a lab schedule, and instructors announce the lab to be done the following week. Thus, students are notified twice about a lab assignment. It is expected that you will preview and prepare for the laboratory (see chapter 7). In college, labs are generally not reviewed in detail, and students are expected to blindly perform the experiments. Many students walk into labs, not even looking at the assigned exercises. That is a definite reading and study problem!

[1] Hein, Best, and Miner, *Foundations of Chemistry in the Laboratory*, Brooks and Cole, 1993.

Another error is the partial reading of the exercises. Laboratory exercises have an introduction to the laboratory exercise. This is followed by the materials, procedure, etc. for the experiment. Since the procedure describes what is to be done in lab, students just read the procedure. They have some idea of what they will be doing mechanically, but no idea what they are doing conceptually. The introduction describes conceptual information and often relates the concepts to the procedures. All too often, students attempt to perform experiments, write reports, or answer questions without having studied the introduction to the laboratory exercise. It is important to read the whole laboratory before coming to the lab!

Sometimes instructors "force" students to read the experiment by giving a pre-lab assignment to be handed in at the beginning of the lab. In this case, the instructors are making the student do the required preparation, but it does not necessarily improve the quality of learning. Some students take the easy way out and copy these pre-lab assignments from lab partners. No reading or studying is done, just like in the old days in high school. This practice might be a temporary survival skill, but is definitely not a good study skill.

Word Lists

Some instructors distribute word lists to be defined and learned. These might or might not be collected and graded. It must be pointed out that these assignments are only guides to terms and do not include everything to be learned. More importantly, you will be expected to be able to *apply* the terms as well as simply *regurgitate* their definitions.

If instructors give these lists to be defined or explained, you should complete the assignment whether they are collected or not. In addition, determine the relationships of the terms to other topics in your class notes. As you do this, you will be comparing and relating the instructor's word list to your lecture and lab notes. This study activity will allow you to check on the quality of your notes and verify the most important things to be learned. However, don't make the mistake of letting these word lists be the only thing you study.

Remember, chapter 9 contains information about terms and symbols.

Study Sheets

Some instructors will compose a series of questions or learning objectives to guide your study. Once again, these might be simply distributed as a study aid, never to be mentioned again. Other instructors use them as worksheets, requiring them to be submitted for grading. Good students tend to include these as only one of the many things they do to learn the course content. They use their notes and the textbook to synthesize an answer. Poorer students tend to fill in the spaces with answers they think are correct and don't refer to books or notes to verify their answers. They are unaware that they don't know the correct information!

As with word lists, study sheets are only a guide to what must be learned. A combination of a word list and study sheet will be a more complete guide than just one or the other.

Recitation Assignments

If you have a recitation, you will be given assignments to complete. Once again, these might be required to be handed in for grading, or they might only be for review. Good students complete assignments. Poor students tend to find excuses why they "just couldn't get the work done." Instructors enjoy working with well-prepared students. Generally they will tolerate unprepared students. However, instructors resent having to explain things to students who are not making any effort and are just "putting in the time."

The assignments are attempts to get students to think critically about the subject material. The general aim is to problem-solve. This type of class is smaller and "gives" the student opportunities to question, analyze, and discuss the information presented in lecture and lab.

Laboratory Work

The syllabus and your instructor will announce the type of written work to be submitted. Some courses require formal laboratory reports on some or all of the labs; others require that the end-of-exercise questions be answered; still others will not require any written work to be submitted. Regardless of the requirements, it is expected that you will learn the information presented in the laboratory exercise. Be sure you know what is expected. Manage your time to do it.

During the lab class, you will follow the procedures described in the manual. Some of the exercises will be observational (for example, observing minerals). Others will be experimental (for example, performing a chemical experiment demonstrating oxidation-reduction). In all cases, you will be expected to observe and record results. The information and notes you record will be needed to write the formal lab report, answer end-of-lab questions, or study for lab tests. When finishing the lab, you should always ask, "Are these results accurate and were they what was expected?" If you don't feel confident with your results, compare your results with results of classmates. If there is still some confusion, you should make an appointment to confer with the instructor.

Your instructor will comment on the lab exercise before, sometimes during, and at the end of the exercise. These comments might:

- point out important parts of the experiment;
- make corrections or clarifications;
- announce common errors;
- emphasize correct results; or
- summarize the procedures and results.

You should **take notes** on what the instructor says. The information will be useful for composing the lab report, answering questions in the manual, or taking tests.

Formal Laboratory Reports Scientists communicate the results of their experiments by writing reports to be published in journals. Writing laboratory reports attempts to mirror the process of writing articles for scientific journals. A format for a formal lab report follows. However, since your instructor may vary this format, be sure to find out what is expected by the instructor.

Cover The cover page should include:

- your name;
- your section number and student number;
- the name of your lab partners;
- the title of exercise/experiment; and
- the date.

Introduction/Purpose This section should be brief and to the point. It should include:

1. An overview of the concepts that relate to the experiment.
2. A few sentences stating the objective(s) of the experiment, the problem to be addressed, and a hypothesis to the problem.

Materials This section could make reference to the list given in the lab manual, or it could be copied directly from the lab manual.

Methods/Procedure Check with your instructor to learn what should be included in this section. Generally, this section is a description of the steps you followed to complete the experiment. Some instructors want the complete procedure copied, others require a paraphrasing, and still others require a summary. Some laboratory manuals number the steps, one after the other. Other manuals describe the steps in paragraph form. Regardless, you should know the sequence of the steps and be able to report on the procedure.

Results/Observations This part of the report is important. It will contain the information you collected during the experiment. This section includes:

- data tables;
- figures of experimental setup or diagrams of things observed;
- graphs;
- notes on personal observations;
- notes or data collected by the class; and
- notes on comments or observations by the instructor.

Discussion/Analysis The discussion/analysis part interprets, explains, and analyzes your results of the experiment. The information provided in your introduction should be related to the experiment and its results. The end-of-experiment questions should help guide what you include in the discussion. It should also include whether or not your results support the hypothesis in the introduction of the report. The essay should demonstrate your understanding of the introductory concepts, the experiment, and its results.

Do not write just one draft of the essay. Most likely your essay will need revision. Critically review your writing, especially for spelling errors, tense constancy, run-on sentences, and repetition of text. If figures and tables are in the results, then you should refer to them in the essay.

It is worth emphasizing that information included in the manual's introduction contains key concepts and relationships that will help you formulate your discussion essay. In addition, the textbook is a valuable reference. It might contain graphs, figures, or explanations that can be used to explain the results of your experiment.

Caution: A common error in writing the analysis is that information is just listed or copied from a reference. The information is not integrated or applied to help explain the results. In addition, an improperly written analysis seems to hope that the instructor will assume that the writer (student) understands how to apply the information. In reality, if information is not well-applied and lots of unrelated information has been included, the instructor will assume that the student cannot apply the information.

Conclusion The conclusion is a brief summary statement of the findings of the experiment or the laboratory.

Errors This section provides you with the opportunity to identify the sources of error in your experiment. Some errors are human errors, like incorrectly measuring the amount of materials or mixing the wrong reagents. Other errors are caused by the limitations of the equipment or possibly the conditions in the laboratory. Balances can measure with only a certain degree of accuracy. Calibrations of equipment and conditions of tools also introduce errors beyond control.

In some cases an "error analysis" is expected to be included in the laboratory report. This might be included in either the error section or the analysis section of the results.

End-of-Lab Questions In some courses, instructors require that end-of-lab questions be answered. Many of these questions will require a sentence or two, a short essay, a figure, a formula, or a mathematical solution. Chapter 13 gives information to help answer these types of questions.

No Written Lab Assignments In some cases, no written assignments are given. The grading for the lab component of the course will be laboratory tests or "practicals." In this type of lab, students have the responsibility to study the lab information on their own. The instructors probably summarize the results at the end of the lab, suggest students answer the end-of-lab questions, and invite students to ask questions during the lab to verify information.

If this is the way your lab is conducted, you should be sure to follow the list on page 42 about what to study for laboratory in chapter 7. Your instructor is depending on your self-direction and self-discipline to study the information covered in lab.

In some courses, the assignments for labs will be a combination of formal lab reports, end-of-exercise questions, and just studying the laboratory exercise. Verify what is expected of you.

Research Papers

Research papers may well be assigned. If so, don't procrastinate about getting started on this assignment. As with other assignments, be sure to define what is expected. If you are not sure how to write a research paper, you should consult your instructor and the writing center at your college. There is nothing wrong with not knowing. However, it is wrong to remain ignorant about requirements and the methods to achieve them.

The following will need to be done to complete a research paper:

1. Choose a topic; adhere to the guidelines set by the instructor.
2. Write a preliminary outline, and submit it to the instructor for review. This might not be required, but it will help you focus on the topic and write the paper. It will also demonstrate that you are academically responsible.
3. Research and collect information about the major topics in your outline. Take notes and begin to create a more detailed outline. Be sure you keep a list of your references in correct bibliographic format.
4. Think about and analyze the information you have collected. Modify and clarify the outline as needed.
5. Write a first draft. Be sure to have a thesis or introductory paragraph (page) followed by pages relating the major points supported by details. Pull things together with a summary or concluding paragraph or page.
6. Do not plagiarize. Use your own words to synthesize and communicate information. Most schools and courses have severe penalties for plagiarism.
7. Edit and proofread the first draft. Make modifications and corrections.
8. Have a study partner critically read and constructively comment on the first draft.
9. Write the final draft that includes the clarifications and corrections suggested by your study partner. Be sure to include footnotes and a bibliography.

Researching and Collecting Information

The process of researching and collecting information can seem overwhelming. A good report will include only pertinent information to fulfill the report's objectives. You will have to extract the information from your resources.

These resources could include your textbook, other textbooks, CD-ROMs, Internet sites, and articles from research journals. For all these resources, it is important to select and organize the information closely related to the objectives of your report. Information gathered from the Internet must be evaluated for its reliability (see chapter 8, Use of Textbooks).

As you do your background research, be certain to keep a list of your sources so that you can properly cite them in your report. Within your report, you must reword all of your gathered information and cite the origin of the information. Be sure to clarify the citation requirements and format with your instructor.

Returned Assignments

When assignments are returned, you should analyze the results. This is an example of an unwritten assignment. If you do this, you will have the chance to learn from your mistakes. Ask and answer the following questions:

1. Is the grade what I expected?
2. Am I satisfied with the grade?
3. What are the mistakes I made in the assignment?
4. What can I learn from these mistakes?
5. What are the comments and corrections made by the instructor?

6. If the grade is poorer than expected, what kind of changes do I have to make in my academic behavior? (Review previous chapters in this book to help you make changes.)

Review

1. College instructors expect you will study; they will not remind you to do so.
2. A variety of study skills and activities should be part of a routine study plan.
3. Textbook, lab manual, and outside readings should be done to reinforce the class work.
4. Your class notes should be related to your readings and other assignments.
5. Word lists and study sheets should guide your study, but they should not be the only things that are studied.
6. Formal laboratory reports have a format that includes a cover page, an introduction, materials, methods, analysis, conclusions, and error sections.
7. In all cases, be sure to verify what work must be handed in for grading, and hand it in on time.
8. Research papers have their own special requirements and take time to complete.
9. Returned assignments should be analyzed to help direct your future work and study.

EXERCISE 12

Name _____

Date _____

Assignments and Reports

....

If you need additional space to answer questions, then use a separate sheet of paper.

1. a. Check which of the following apply to you.

_____ I initiate my own studying to learn the information presented each week.

_____ I depend on the teacher to direct my study.

_____ I expect instructors to review the information that will be on the tests.

_____ I use my notes to help direct and focus my readings and study of the textbook.

_____ I generate learning objectives to direct my study.

_____ I create questions to test myself and monitor what I have learned.

_____ I skim through the textbook but depend on my notes to guide what I study and learn.

_____ I depend on reading the textbook because I take poor notes.

_____ I complete written assignments on time.

_____ I know the format for a formal lab report.

_____ I realize that I must relate information in the lab manual and textbook to the results of my experiments.

_____ I generally procrastinate and do things at the last minute.

_____ I consider the time spent studying each week a good way to learn and prepare for tests.

_____ I manage my time so that I can complete the assignments given to me and to preview and review the information presented in classes.

b. List three actions you can take to improve your study habits.

_____

_____

_____

2. a. List any adjustment problems you have had in reference to your academic behavior.

_____

_____

_____

_____

b. List an action you can take to resolve each problem.

_____

_____

_____

_____

c. Share and discuss your answers in a small group of classmates.

d. Make a group list of problems and record these on the board.

e. Have a discussion about how these problems might be resolved.

3. List routine, self-directed, and self-disciplined assignments that you will do for your classes (for example, reading, flash cards).

Lecture

_____

_____

_____

Laboratory

_____

_____

_____

93

Recitation

_____

_____

_____

(Also see Appendix B)

4. Describe the format for a formal laboratory report.

_____

_____

_____

_____

_____

_____

_____

_____

_____

_____

5. Which part of the lab report will be the most challenging to write? Why?

_____

_____

_____

6. What is the difference between reading and studying?

_____

_____

_____

_____

_____

7. a. List the types of assignments covered in this chapter.

_____

_____

_____

_____

b. For each type of assignment, describe why personal self-direction and self-discipline are important.

_____

_____

_____

_____

_____

_____

_____

8. Define the term "plagiarism."

_____

_____

_____

CHAPTER 13

Answering Essay and Math-Based Problems

Objectives

When you have read this chapter, you will be able to answer these questions:

1. Is there a system or process to help me solve problems?
2. What are some guidelines to writing a good essay answer?
3. What are some guidelines to help me solve math-based problems?

Assignments and tests are directed toward solving problems or answering questions. If you develop a systematic approach to problem-solving and practice that approach, you will improve your problem-solving skills. If you practice writing essays or solving math-based problems, you will improve your skill of essay-writing or problem-solving. Thus, if assignments are given, you should complete the assignments to the best of your ability. This effort will give you practice and help prepare you for questions on tests.

The Problem-Solving Process

A question poses a problem that can be answered with a written answer or a mathematical calculation. Written answers can be as short as one sentence or as long as several pages. Writing an essay or solving a math-based problem requires that you (1) **analyze** the problem, (2) **plan** a solution, (3) **answer** the problem according to a plan, and (4) **evaluate** the solution (figure 13.1).

What do authors of science textbooks say about the characteristics of students who are good problem-solvers? Bodner and Pardue list the following characteristics in their textbook.[1]

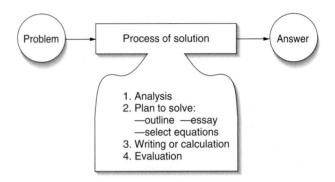

Figure 13.1 The process of solving a problem has four components.

Good problem-solvers:

- believe they can solve almost any problem if they work long enough;
- read carefully and re-read a problem until they understand what information is given and what they are asked to solve for;
- break problems into small steps that they solve one at a time;
- organize their work so they don't lose sight of what they've accomplished and can follow the steps they've taken so far;
- check their work, not only at the end of the problem, but at various points along the way;
- build models or representations of the problems that can take the form of a list of relevant information, a picture, or a concrete example;
- try to solve a simpler, related problem when faced with a problem they can't solve; and
- try out several approaches to a problem until they are successful.

Answering any question, whether it be on an assignment or on a test, involves a problem-solving procedure. Preparation for a test, types of tests, and test-taking will be

---

[1] Bodner and Pardue, *Chemistry: An Experimental Science,* Wiley, 1989.

Table 13.1 Characteristics of Essay Tests Source: The definitions of answers needed and examples of specific terms are adapted from Jason Millman and Walter Pauk, *How to Take Tests* (New York: McGraw-Hill, 1969), pp. 152–57. Used with permission. The form of the table is taken from "The Vocabulary of Test-Taking." Reprinted by permission from the 1975–76 issue of *Nutshell.* Copyright © 1975 by 13–30 Corporation.

General Category	Answer Needed	Examples of Specific Terms
Identification	Present the bare facts: a date, name, or phrase; in short, provide a concise answer.	Cite, define, enumerate, give, identify, indicate, list, mention, state
Description	Tell about a specific topic with a certain amount of detail.	Describe, discuss, review, summarize, diagram, illustrate, sketch, develop, outline, trace
Relation	Describe the similarities, differences, or associations between two or more subjects.	Analyze, compare, contrast, differentiate, distinguish, relate
Demonstration	Show (don't state) why something is true or false. Put forth logical evidence or arguments to support a specific statement.	Demonstrate, explain why, justify, prove, show, support
Evaluation	Give your opinion or judgment on a subject, plus justify and support it. Also, if your opinion can be challenged, be sure to present both sides.	Assess, comment, criticize, evaluate, interpret, propose

discussed in the next chapter. Nevertheless, what is said in this chapter applies to solving objective questions also.

Writing One-Sentence Answers

Some questions require an answer that is just a sentence or two. This sounds simple but is often difficult to achieve. Common errors students make are that they:

- do not analyze the question correctly and make up their own question;
- do not plan to use the terms and vocabulary of the question;
- do not write complete sentences or complete thoughts; or
- do not recognize that what they have written does not make sense.

A plan or technique to successfully answer this type of question is to include words and phrases of the question in the answer. If you apply this technique, you will avoid the above errors. You must also have the correct information to complete the sentence. If you don't have the completing information, you will make a lack of knowledge error rather than a grammatical error. The following are three examples:

1. What is the effect of bases on the color of red litmus? The one-sentence answer should begin: The effect of bases on the color of red litmus is . . .
2. What are protein polymers made of? The one-sentence answer should begin: Protein polymers are made of . . .
3. What is the function of a mitochondrion? The one-sentence answer should begin: The function of the mitochondrion is . . .

Writing Essay Answers

Writing essay answers is more complex. The following **outlines** how the **problem-solving process** applies to writing an **essay.** All essay tests use certain test terms or qualifier words that indicate the type of answer that is needed. It is important that these terms be understood to help you analyze and define the question. Table 13.1 summarizes the general categories, gives the type of answer needed, and gives examples of specific terms used in essay questions. It is important that your essay answers are appropriate and contain only essential information. It's common for students to write more than is necessary, including run-on thoughts and streams of information that don't answer the question. This approach will not gain you points on an exam (it might actually cause you to lose marks) and will not help you to learn science. When answering an essay question you should:

I. **Analyze the Problem**
    A. Read the question carefully; underline key terms.
    B. Define the key terms and think about their relationships.
    C. Identify the test terms such as "compare," "differentiate," or "interpret."
    D. Identify the problem to be answered and determine if the question is divided into "sub-questions."
        1. Number the sub-questions to help identify them.
II. **Plan a Solution**
    A. Think about the relationships and information in the question.
    B. Jot down thoughts, facts, figures, and symbols in a "memory dump."

C. Organize an outline of the "memory dump" information to answer the question.
    1. List major topics.
    2. Support major topics with details or examples.
    3. Be sure scientific terms in question are included in the outline.
    4. If appropriate, sketch a simple diagram to illustrate the essay.

III. **Answer the Question**
    A. The essay answer should be written in three parts.
        1. Begin the essay with a thesis sentence, sentences, or paragraph. This beginning should restate the problem and indicate how you will answer the question. Use the same test and scientific terms that are in the question.
            a. A question from a geology book states: "Describe two sources of heat causing metamorphism." A thesis sentence might read: "Two sources of heat causing metamorphism are. . . ."
        2. The body of the essay contains the supporting information. The detail of the outline should make this easy to write.
            a. Write a paragraph about each major point and support it with other detailed information.
            b. Don't add a lot of unnecessary words to embellish the essay.
            c. Use the terminology of science and the terms in the question.
            d. If a diagram is appropriate, include one, but bear in mind that a diagram by itself is not an acceptable answer to an essay question.
            e. Leave a few lines between paragraphs just in case you might want to add something after you have reread the answer.
            f. Write a general answer even if you don't know the specifics.
        3. The concluding sentence, sentences, or paragraph should summarize how you have answered the problem. "Thus, the heat to form metamorphic rock can come from the geothermal gradient and by plutonic activity. . . ."

IV. **Evaluate the Solution**
    A. Re-read the essay. Compare the content to the outline, making sure nothing was omitted. If so, use the lines between the paragraphs to include the omissions.

B. Does the essay have the following characteristics?
    1. Accuracy
    2. Relevancy
    3. Organized content and style
    4. Completion, all parts answered
    5. Correct grammar and spelling
    6. Neatness

Many of the questions in the exercises of this book require a written answer. Some of these answers are one sentence, others are short essays. Consult your instructor to check on the style of writing answers to these types of questions. This type of analysis should help you answer similar questions.

Two examples of essay questions are as follows:

1. Describe what is meant by a balanced chemical equation. Give two examples to demonstrate this concept. Notice that:
    • "Describe" is the test or qualifier term.
    • "Balanced chemical equation" is the key scientific term.
    • The question is made of two parts.
    • The second part asks for two examples.

A planned outline could be created from the information on page 78. From this plan, an essay could be written utilizing the above four points.

2. Differentiate between dehydration synthesis and hydrolysis. Give two examples of specific molecules involved in these reactions. Note that:
    • "Differentiate" is the test or qualifier term. Refer to table 13.1 to determine its meaning and what should be included in the answer.
    • "Dehydration synthesis," "hydrolysis," "molecules," and "reactions" are scientific terms.
    • The question is made of two parts.
    • The second part asks for "specific examples of molecules."

Information about these questions can be found in figure 8.1 and on pages 66 and 79. An outline and essay could be written to differentiate, use the scientific terms, answer both parts, and give specific examples.

Solving Math-Based Problems

Math questions could be given as assignments for lecture, lab, or recitation classes. Instructors will announce whether they will be part of tests. However, if you are taking chemistry or physics, there will be math-based problems on all your tests. The following should guide how you answer math-based problems. Also refer to the flow-chart on page 86.

I.   **Analyze the Problem**
   A.   Read the question carefully, underlining key terms and units.
   B.   Define the key terms and think about their relationships.
   C.   Identify any words or phrases that give a clue to a solution, such as "sum of," "equal to," or "products of."
   D.   Identify the problem to be answered and determine if the question is divided into "sub-questions."
   E.   Determine what principles or concepts the problem involves.
   F.   Display the problem.
     1.   Write down what is known (given) in the question and what is unknown (wanted).
     2.   Draw a simple diagram and label what is known and unknown.
     3.   Identify whether or not part of the solution requires the conversion of units. Do this before you begin to solve the problem.

II. **Plan a Solution**
   A.   Think about the relationships and information in the question.
   B.   Decide which equations will relate the known and unknown factors. Realize that one unknown in a problem will require one equation in the solution. Two unknowns will require two equations.
   C.   If more than one equation is required, be sure to plan the sequence of how to apply the equations. If you can integrate the equations, do so before you calculate the answer.

III. **Answer the Question**
   A.   Insert the known values into the equation.
     1.   Check the units of the known variables. The units of the given information should fit into the equation.
       a.   If mass is called for, make sure you don't try to insert a unit for a rate of some sort.

   B.   Do the calculations to solve for the unknowns.
   C.   When the calculation is complete, be sure to label the answer, designating units.

IV. **Evaluate the Solution**
   A.   Check to see if the answer is reasonable.
   B.   Check your calculations for mathematical errors. Decimal marks have a tendency to be misplaced.
   C.   Check to see that what was wanted in the analysis was given in the solution.
   D.   Check to see if you have the right number of significant figures. The most common errors in mathematical problems involve the following:
     1.   incorrect units or improper unit conversion;
     2.   use of the wrong equation;
     3.   solving for the wrong variable;
     4.   mistake in the equation or calculation;
     5.   selection of improper information from a complex problem; and
     6.   lack of ability to relate a sequence of mathematic expressions to a problem.

Review

1.   Problem-solving is a skill that can be improved with practice.
2.   Questions pose problems that can be answered with a written or mathematical solution.
3.   Problem-solving requires that you analyze the problem, plan a solution, answer the problem according to the plan, and evaluate the answer.
4.   A sequence of steps should be adopted to compose written answers and solve mathematical problems.

EXERCISE 13

Answering Essay and Math-Based Problems

....

1. a. Check which of the following apply to you:

_____ I generally "dive right into" answering a question by writing an essay or solving a math problem.
_____ I take time to analyze a problem, plan a solution, and solve the problem according to the plan.
_____ I evaluate the problems I solve by proofreading my work.
_____ I create outlines before I write essays.
_____ I always lose points on essay questions.
_____ I record what is known and unknown and create a diagram when I try to solve a math-based problem.
_____ I practice solving problems by answering end-of-chapter questions in the textbook and lab manual.
_____ I have trouble solving math problems.
_____ I always check the units when doing math-based problems.
_____ I have trouble writing essays because I'm weak in English.
_____ I analyze the comments instructors make on the essays they return.
_____ I check the mistakes I make on math problems to help remind myself not to make the same errors.
_____ I practice solving math problems given in the textbook.
_____ I answer any sample questions the instructor might distribute before a test.

1. b. List three actions you can take to improve your problem-solving skills.

_____

_____

_____

2. Eight characteristics were given to characterize good problem-solvers in this chapter. List which of these characteristics apply to you.

_____

_____

_____

_____

_____

3. List the four things that are involved in the process of problem-solving.

_____

_____

_____

_____

4. Outlines are given in this chapter to suggest a system to solve essay and math-based questions. On a separate sheet of paper, write an essay describing how to answer either an essay or math-based question. (Feel free to use the outline given in this chapter.)

5. a. Underline the key words in questions 2, 3, and 4.

5. b. Which question(s) ask you to list things?

_____

5. c. How many essays must you write to answer question 4?

_____

6. Underline the key words that are in the following questions:

6. a. Describe how a triglyceride is synthesized.

6. b. Explain what is meant by a "balanced chemical equation" and give an example.

6. c. Explain the difference between a triglyceride and a protein.

7. Write the beginning phrases for the thesis sentences to each of the questions in 6.

7. a. _____

7. b. _____

7. c. _____

8. Which of the three questions in 6 has two parts?

_____

9. Describe a carbon atom. Indicate its potential to bond with other atoms.

   The following are two answers to the above question. The question is worth 10 points on a test. Evaluate each answer and give it a grade.

Answer A

Grade: _____

A carbon atom has a core and an outside. The center is made of protons and neurons. The outside has things moving around fast, like a cloud, and these things react with other things. I think this bonds them together. The protons and the things on the outside balance each other out.

Answer B

Grade: _____

A carbon atom is made of two basic parts, the nucleus and shell (orbits), or electron cloud. The characteristics of the shell(s) determine the bonding potential with other atoms.
   The nucleus of carbon has six protons that are positively charged. There are also six neutrons in the nucleus.
   There are two shells that have electrons in them. These are negatively charged. The first shell has two electrons, the second shell four electrons. All electrons are moving around the nucleus very fast.
   The second shell really has room for eight electrons. Four more electrons can fit into this outer orbit for short periods of time. Thus atoms that could share their electrons with carbon could form a bond with carbon. For example, four hydrogen atoms, each with one electron, will chemically bond to carbon.
   Thus a carbon atom has two parts, the nucleus and electron shells. The outer shell determines its bonding potential.

10. What are the deficiencies of the poor essay and the proficiencies of the good essay? Refer to the outline of a good essay answer given in this chapter (p. 96).

Deficiencies	Proficiencies
_____	_____
_____	_____
_____	_____
_____	_____
_____	_____
_____	_____

_____ _____

_____ _____

11. A math-based question in chemistry reads "Calculate the moles and grams of lead(II) nitrate present in 10.0 mL of 0.50 M $Pb(NO_3)_2$ solution."

11. a. Underline the key words in the question.

11. b. List what is known: _____

   _____

   _____

11. c. List what is unknown and to be found:

   _____

   _____

11. d. What are the units in this question? _____

   _____

   _____

12. A physics problem reads as follows: A metal ball is dropped from the top of the Empire State Building. The ball takes 8.9 seconds to reach the sidewalk. Determine (a) the velocity of the ball when it hit the sidewalk, (b) the average velocity of the ball, and (c) the distance the ball fell from where it was dropped to the sidewalk.

12. a. This question is in fact made of _____ parts.

12. b. What is given: _____

   _____

12. c. What must be found in each part of the question?

   _____

   _____

12. d. Draw a diagram of the problem. Include in the diagram a picture of a building, the sidewalk, a falling metal ball, and an arrow to show the path of the ball. Then label the time it took to fall, where its final velocity is reached, where it begins to increase its velocity in free fall, and the distance the ball falls.

12. e. Why is it important to understand the difference between the beginning velocity, average velocity, and final velocity?

12. f. How many equations would you hypothesize are needed to solve this problem? _____

13. Imagine that the above physics question was on a test. Describe the information, in a general sort of way, that might have been covered in the physics classes or recitations prior to the test. In your description, try to predict the topics covered, the types of problems given as assignments, and the types of equations presented by the instructor.

_____

_____

_____

14. Three different answers are given to the same test question. Pick the best answer and indicate the type of error(s) in each answer.

Question: Four words are listed. One word in the group can be considered different. Underline the different word and explain how it is different. Explain what links the other three words.

carbon, water, hydrogen, oxygen

Answer 1:

carbon, <u>water</u>, hydrogen, oxygen

Water is a liquid; the rest are gas.

Answer 2:

carbon, <u>water</u>, hydrogen, oxygen

Each one, carbon, hydrogen, and oxygen, are gases. Water is a combination of the three.

Answer 3:

carbon, <u>water</u>, hydrogen, oxygen

Water is a compound composed of two elements. The other three are elements.

_____

_____

_____

_____

_____

15. Differentiate between dehydration synthesis and hydrolysis. Include in your discussion the role of enzymes and an example of each type of reaction, and include simple diagrams to help illustrate synthesis or hydrolysis.

15. a. List the key terms:

_____

_____

15. b. How many sub-parts are there to this question?

_____

15. c. What does each part of the question ask you to do?

_____

_____

_____

_____

15. d. What does the word "differentiate" mean? (Refer to table 13.1.)

_____

_____

_____

16. Write a question that would require an essay answer to the following:

16. a. Figure 9.5.

_____

_____

_____

16. b. Map, figure 9.6.

_____

_____

_____

16. c. Table 13.1

_____

_____

_____

CHAPTER 14

Tests

Objectives

When you have read this chapter, you will be able to answer these questions:

1. How can I prepare for a test?
2. What kinds of questions can I expect on tests?
3. What skills can help me get better grades?

"The test is coming, the test is coming!" Yes, indeed, **tests** will be given to **evaluate** your **knowledge** of scientific information. Check the course syllabus to determine the number, types, and dates of tests. Quizzes (short tests) might be given in lab or recitation. You must **prepare** for, **take,** and **analyze** the **results** of each test.

Just about everyone feels a certain anxiety about taking tests. If you prepare for tests, apply good test-taking skills, and learn from your mistakes, you should become a confident test-taker. If your anxiety levels are high even before you begin the science course, you will have difficulty with tests. Seek help to learn how to reduce your anxiety about test-taking. (See chapter 1.) As you prepare for tests:

- Concentrate on the information to be learned.
- Don't dwell on your imagined inability to learn.
- Don't connect your grades with self-worth.
- Don't create negative feelings about yourself or the course.
- Concentrate on your efforts and attitudes, and don't worry about other people's efforts and attitudes.
- Don't magnify your fears.
- Control and organize your actions.

Some students smile with satisfaction and walk lighter on their feet because they are proud of the good grades they have earned. They think to themselves, "Yes, it was worth spending all that time studying; it paid off!" Some students who receive low grades say, "I didn't know what to study" or 'The '#@*^' instructor never gave us a review or even told us what to study" or "Oh, well, I've always done bad on tests, especially in science." Instructors present information and establish standards in the course. It is your job as a student to make the effort to meet those standards (figure 14.1).

Preparation

Start to study for tests on the first day of class. You can do this by:

- establishing a cycle of weekly study,
- defining learning objectives and accomplishing them,
- organizing weekly summaries,
- studying with classmates,
- completing assignments,
- correcting and enriching class notes,
- generating your own test questions, and
- testing yourself to monitor your learning.

Review other suggestions given in this book to guide how you can prepare for tests. Remember, you should be studying between 6 and 12 hours each week! Frequent reviews allow you to comprehend and remember the information. They also allow you to relearn forgotten information. The final review before a test just sharpens what you have already learned.

As you prepare for tests, you must be able to **recognize** and **recall** information. Superficial study will allow you to recognize information and possibly answer multiple-choice, true/false, and matching questions. These types of questions give all of the information and the right answer in the question. An ability to recognize information and good test-taking skills might allow you to do well on these tests. However, fill-in, essay, and math-based tests require you to recall and apply information. Thorough preparation for tests should enable you to recognize as well as recall information.

Cramming or some other last-minute study system isn't worth talking about.

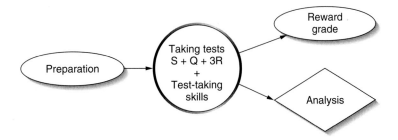

Figure 14.1 You must prepare for tests, take tests, analyze the results, and hopefully be rewarded for your efforts.

The Week before the Test

Begin to focus on a test five to seven days before the test is to be given. Find out as much about the test as possible:

- How many questions will be asked?
- What types of questions will be asked?
- How long will you have to finish the test?
- Do you need to bring pens, pencils, or special answer sheets?
- What are the topics to be covered on the test?
- Will you be able to keep the questions for review purposes?
- Are sample tests available in the library or website?

The answers to these questions will help you prepare for and take the test. For instance, will the test be all **objective questions** or a combination of objective and **subjective questions?** Will there be 40 or 60 objective questions? Will you have 50 or 75 minutes to finish the test? This information is important to know.

Objective questions tend to be about specific and detailed information. Studying for that type of question is different from preparing for essay or math-based questions. If essay questions are to be answered, you must prepare by studying the overall approach to the course content. You should be able to apply concepts and principles and support them with examples in written statements. You should try to predict what essays might be asked. Listen for clues and record these in your notes, and put exclamation marks next to the clues. If an instructor spends a week enthusiastically discussing a topic, you should be able to write an essay about what has been discussed. Obviously if an instructor gives you sample questions, you should prepare answers for all of them.

If mathematical problem-solving is to be on the test, you must practice, practice, and practice solving problems as you prepare for the test. Be sure you can solve examples of each type of problem discussed in class. Remember, these types of tests apply the principles studied in class and in the textbook.

To **prepare** for a test, you should do the following:

1. Study the weekly summaries you have organized.
2. Review your corrected and enriched notes and identify areas you feel uncertain about. Concentrate on these unfamiliar areas.
3. Survey pertinent parts of the book (those areas covered in lecture).
4. Practice solving problems.
5. Continue to work with flash cards or vocabulary-definition lists.
6. Test yourself with questions from the Questions column of your notes and the questions that appear in the textbook.
7. If sample tests are available, take these at least two days before the test. Study the content of the questions you got wrong. Be sure you know the right answers. Don't just memorize the right answers; know why they are right.
8. Meet with a study group at least two or three days before the test.
9. Be able to explain all figures used by your instructor.

The Day before the Test

The final review should be done the day before the test. Visualize and recall the information presented week-by-week to you. After trying to recall this information, you should once again review the summaries, survey your notes, reproduce figures, skim the textbook, and review the questions. Take a break. Then skim your questions one last time and try to predict which questions will be on the test. Make up your own test and answer it.

Get a good night's sleep! Sleep on what you have learned.

Test Types

Tests can include questions that are **objective** or **subjective** or that require **mathematical** calculations. Any test is meant to measure your knowledge and understanding of information. As indicated in chapter 13, all questions require you to analyze, plan or organize a solution, give an answer, and evaluate the answer in a problem-solving process. Tests may be given on computers, in class on paper, or as take-home exams.

Objective Test Questions

Objective test questions require that you select or give a specific answer to a question. You must be able to recognize and recall what you have learned to select the correct answers from the choices given. These types of tests are called objective because there is only one correct answer; there is no "partial answer." Occasionally, questions are poorly constructed.

Multiple-Choice Multiple-choice questions can be considered to be a series of true/false questions packaged into a single question. These tests require you to select from four or five possible answers. A question might have answers like "all of the above," "none of the above," or "b and d are correct." Questions with this type of answer relate information from different parts of the course. They require students to apply as well as recognize and recall information. Read these types of questions and choices with particular care.

True/False In a true/false format, you have to select whether a statement is one of two choices: true or false.

Matching These tests require the matching of terms, phrases, or sentences in one column to terms or phrases in another column. If you know all the key words (test words), you should do well on this type of test.

Identification and Fill-In These types of questions require you to recall specific words or phrases and to record these answers on the answer sheet. These questions are used most frequently on laboratory tests.

Laboratory Tests

These types of tests are also called lab "practicals." The type of test given on the content of laboratory varies. These tests focus not only on the scientific information but also on the materials, procedures, and results of laboratory experiments. Review the list of things to learn in laboratories given in chapter 7.

In some cases, laboratory test questions are spread around the room at question stations. Student will move from one question station to the next. Questions are asked about specimens, equipment, chemical reagents, graphs, or whatever else is on display at the station. These laboratory tests might be timed tests, one minute or so per question station.

It should be emphasized that you should take good notes on all aspects of the laboratory work. Be sure to record hints and reviews given by the instructor and have complete diagrams from labs. Review diagrams that are drawn during labs. These notes should be enhanced during

study sessions to create a study tool to help you learn laboratory information and prepare you for tests.

Subjective Test Questions

Subjective questions require answers ranging from a sentence to an essay. These questions not only test your ability to recall information but to organize and express this information in a written format. When answers are graded, partial credit can be earned. The instructor will be looking for specific information, will evaluate how information is related and organized, and will be influenced by how well the answer is written. Review the information in chapters 12 and 13 about written assignments.

Math-Based Problems

Questions that require you to calculate answers require knowledge of mathematics. Equations can be memorized, but only practice of problem-solving will enable you to apply these equations to math-based questions on tests. There is a correct answer to a math-based question. However, the thought processes as displayed in the presentation of your calculations are evaluated. Partial credit is given for parts that display correct thinking even though the final answer might be wrong. Review the information in chapter 13 about math-based problem-solving.

Take-Home Tests

This type of test might seem easy at first glance but actually is often more challenging. A specific due date is set for the return of the test. The types of questions generally require students to compare and relate different aspects of the course. Well-composed answers are expected.

Taking Tests

Taking a test allows you to apply what you have learned. Use the principles of S + Q + 3R and good test-taking skills. Proper preparation and an analysis of returned tests (see figure 14.1) are also important parts of the entire process. The analysis of test results will be discussed in chapter 15. The approach to answering essay and math problem-solving questions was discussed in chapter 13. When taking a test, you should:

1. Arrive for the test on time, but avoid getting suckered into a "Did you study this, did you study that?" discussion with other students. Be confident in your own preparation for the test. Remember, you are in charge, you have taken control of your studying and learning.

2. Accept the test. Listen attentively to all instructions and corrections given by your

instructor. Make changes on your test then and there.

3. **S + Q + 3R** the test

    a. **S—Survey**

    Begin the test by surveying the entire test. Take note of the following:

    - directions and time allowed for test;
    - number and types of questions;
    - sequence of test (Does the test follow the lecture sequence, or are the topics mixed up?); and
    - value of each part of the test.

    b. **Q—Questions**

    The questions are generated for you, but as you begin to read the instructor's questions, realize that you will want to ask yourself other questions that will help lead you to the correct answer.

    c. **R—Read**

    Read slowly and carefully; every word counts. You should:

    - read and follow all the directions;
    - read each question, being sure not to change the meaning of the question;
    - use your pencil and guide yourself through the question;
    - underline the key words to slow the pace of your reading and identify important terms in the question;
    - read all parts of the question;
    - read and answer the easy questions first; and
    - go back to the more difficult questions later for still another reading and analysis.

    d. **R—Recall**

    After reading each question, recall information from your study about that question. You should:

    - think of the answer;
    - select the correct answer if it is a subjective type question or record the answer if it is a fill-in;
    - not dwell or linger on it if a question is difficult and you can't recall the information or can't think of the answer. Mark the question and return to it later. Your mind might subconsciously search for the answer to it. Take time to recall or possibly recite information about the topic of the difficult question.

    e. **R—Review**

    After answering all the questions, check the time, then:

    - if you have the time to relax a minute, do so;
    - return to the test and review all questions if there is time;

- don't change an answer without due cause; and
- check to see if answers are in their proper place and all questions are answered.

Finishing the Test

Before handing in the paper, make sure your name and identification number are on the answer sheet. If it is a computer-graded answer sheet, make sure you have erased all stray pencil marks. Hand in the test. You will probably talk about the test with other students after the exam is done. Try not to develop a firm result of your grade in your mind. Wait until the tests are returned.

Test-Taking Hints

1. If you have any questions about the directions or the test questions, you should ask for clarification.
2. After surveying the test, plan how you will spend your time. Determine how much time to spend on each part of the test or on each essay or math-based problem; keep track of time, but try not to worry about it.
3. *Watch* for the test or *qualifier words* in the questions.
4. *Make* an *educated guess* to the answer if you are not penalized for guessing.
5. Information in one question might contain the answer to another question.
6. *Mark* up the test *to help* yourself identify and clarify the question.
7. Answer easy questions first to build confidence.
8. Take the questions for what they are; *don't read hidden meanings* into the questions.
9. *Mark* questions *you skip* so you don't have to search for them. Don't forget to skip the question on the answer sheet.
10. Don't let the pace of others influence your test-taking; use all the time given to you if you need it.
11. *Watch for word clues.* Statements that express absolute quantity (all, always, exactly, none, never) are usually false.
12. *Keep checking* that you are placing the answers in the proper location on the answer sheet.
13. Bring sharpened pencils and extra pens to the test.

Multiple-Choice Questions

1. After reading each question, try to think of the answer.
2. Read all the choices to find the correct answer; don't stop at the one you first think is correct.
3. Note whether the choices include a combination of answers. Be doubly careful to read all choices.

4. It is helpful to cross out choices that are obviously wrong, leaving only the possible answers to think about.
5. Have a good reason to eliminate a possible answer.
6. Correct answers will read as a true statement as you match the question and the answer. (All other choices will make the question into a false statement.)
7. If you don't know the answer, guess at it after eliminating the obviously wrong answers.
8. Recognition might help you answer the question, but recalling information is a better basis for making choices.

True/False Questions

1. If any part of the statement is false, the whole statement is false. Don't be tempted to answer true because it sounds good; all words and phrases must "fit" for the statement to be true.
2. Watch for qualifier words like "all," "some," "normally," "sum of," or "must." These give clues as to whether it is true or false.

Matching Questions

1. Read each column.
2. Re-read each choice in the left-hand column and recall information about the statement or word.
3. Think of possible answers and look for the answer in the right-hand column.
4. Cross out used choices if the directions say that possible answers can be used only once.
5. Remember to do the easy ones first, the more difficult ones last.

Subjective Test

1. Plan your time carefully.
2. Create outlines from "memory dumps" to guide the writing of essays. (Refer to chapter 13.)
3. Try to learn the type and form of answers the instructor likes.
4. An outline or diagram is better than leaving a question not answered.
5. Be sure you answer all the parts of the question.
6. Consider labelling or numbering parts of your essay.

Review

1. Preparation for tests should start on the first day of class.
2. Study sessions should be spent identifying what must be learned, organizing the information to be learned, and learning the information. This learned information will then be applied on tests.
3. If you are test-anxious, you should take measures to reduce your anxiety level.
4. You should find out as much about the test as possible beforehand. Do not be surprised by the format of the test.
5. Begin reviewing for a specific test one week before.
6. Utilize the study skills and exercises that have been discussed in this book to guide your preparation for tests.
7. Test questions will be objective, subjective, or math-based.
8. Essay questions require a "memory dump," organizing an outline, and composing a well-written answer.
9. Practicing solving math-based questions is an important way to prepare for math-based tests.
10. Successful studying for tests depends upon establishing a cycle of weekly study.
11. Cramming is a slipshod method of learning.
12. Study and review summary charts and maps, class notes, text, figures, and questions.
13. Review difficult areas first.
14. Learning depends upon frequent reviewing to enable you to relearn forgotten information.
15. Frequent study sessions spread the preparation out, making the study easier.
16. When taking tests, use the principles of S + Q + 3R and good test-taking skills.
17. Read each question slowly, carefully, and completely. Visualize notes, figures, and summaries to help you answer the question.
18. Review the test before handing it in, proofread the answers, and check to see if you have answered everything.

EXERCISE 14

Name _____

Date _____

Tests

....

1. a. Check which of the following applies to you. Review this checklist after the first test. Re-evaluate the way you study.

_____ I study 6 to 12 hours each week.
_____ I direct and organize my study to prepare for future tests.
_____ I feel very anxious about tests.
_____ I have trouble identifying what I need to study.
_____ I have trouble concentrating as I study.
_____ I generally feel rushed to finish a test.
_____ I am confused by all of the choices on a multiple-choice test.
_____ I practice solving math-based problems to prepare for a test.
_____ I frequently feel I have not studied enough for a test.
_____ I always seem to study the wrong information.
_____ I guess a lot on objective tests.
_____ I do the easy questions first, then really concentrate on the difficult questions.
_____ I worry about what's going to happen if I fail the test.
_____ I depend on the clues in the question to answer the question rather than recalling the correct answer.

1. b. Review this checklist after the first test, and circle those actions that you can change to improve test-taking skills.

1. c. For each action, specify one way to improve upon your test-taking skills and confidence.

_____

_____

_____

_____

_____

2. Check which of the following test-taking skills you use. Reevaluate yourself after the first test. Did you use these skills?

_____ Practice some form of relaxation before the test.
_____ Come to the test on time.
_____ Bring extra pencils and pens to the test.
_____ Listen attentively to verbal additions and corrections.
_____ Make corrections on the test before starting to take the test.
_____ Read the directions.
_____ Survey the entire test before taking the test.
_____ Use a pencil as you read the test to guide attention to the question and the key words in the question.
_____ Underline key words and phrases to slow yourself down and to read the questions more carefully.
_____ Analyze the relationships expressed in the questions.
_____ Look for word clues.
_____ Visualize or recall notes, figures, and summaries to help answer questions.
_____ Try to think of an answer to a question before reading the choices given in the question.
_____ Read all choices in a multiple-choice question.
_____ Try to find answers to unknown questions on other parts of the test.
_____ Eliminate obviously wrong answers in multiple-choice questions before selecting the best choice.
_____ Review the whole test if enough time is available.
_____ Analyze questions that were wrong on a returned test.
_____ Try to identify why you gave the wrong answers on a marked test.
_____ Keep old tests for review during preparation for the final examination.

3. Here are some examples of test questions that appear on science tests. (Answers are in Appendix B.)

Practice the beginning of test-taking by identifying the key words in the following sample questions. This activity helps you to practice the first "R" of the S + Q + 3R technique to analyze and identify important terms in the questions.

Example 1—Oceanography

True/False: The foreshore is that portion of a beach that is never covered by water even at high tide.

Multiple-Choice: Which of the following is in the middle layer of a stratified body of water?

a. epilimnion

b. hypolimnion

c. thermocline

d. homothermous

e. lithosphere

Example 2—Meteorology

True/False: A bar is a unit of pressure equal to 1,000 millibars, or 29.53 inches of mercury.

Multiple-Choice: Thunder is

a. the clash between two colliding weather fronts.

b. the side effects of the tremendous energy released by a discharge of lightning.

c. caused by cumulonimbus clouds surging upward against a cold air mass.

d. the results of the aftereffects of a tornado.

e. none of the above.

Example 3—Geology

True/False: In undisturbed layers of sedimentary rock, the oldest rock is on top, the youngest on the bottom.

Multiple-Choice: Quartzite is the metamorphic product of

a. quartz sandstone.

b. granite.

c. limestone.

d. shale.

e. rhyolite.

Example 4—Biology

True/False: Plants are multicellular, eukaryotic and autotrophic, and have cell walls.

Multiple-Choice: Hormones and pheromones are organic molecules that function to

a. deliver some sort of message.

b. provide the organism with energy.

c. protect the organism from pathogens.

d. act as molecular building blocks.

e. serve as cell receptors.

A protein is an example of

a. a polymer.

b. a molecule made of amino acids.

c. a triglyceride, made of three fatty acids.

d. a and b.

e. a, b, and c.

Example 5—Chemistry

True/False: In forming an ionic bond, each of the reacting atoms shares two or more electrons between the atoms.

Fill-in-the-Blank: Fill in the correct answer.

What is the molarity of chloride ions in these solutions?

a. 0.10 M NaCl _____

b. 0.10 M KCl _____

c. 0.10 M $BaCl_2$ _____

d. 0.10 M $AlCl_3$ _____

4. This question will help you to practice parts of the S + Q + 3R technique for taking tests. Read the following chemistry question and then answer questions a–e.

Name three indicators of acids and bases. Describe the colors they exhibit at different pHs.

4. a. What kind of question is this?

_____

4. b. How many parts are there to this question?

_____

4. c. List the key words in this question.

_____

_____

_____

4. d. Which words in the question do you consider as test or qualifier words. (Remember, these are words that tell you what to do.)

_____

4. e.   How many paragraphs would you write to answer the second part of the question?

_____

5.   This question will help you to practice parts of the S + Q + 3R technique for taking tests. Read the following chemistry question and then answer questions a–d.

How many milliliters of a 2M HC1 solution would just neutralize 25 milliliters of a 1M NaOH solution?

5. a.   What kind of question is this?

_____

5. b.   What units must be considered as you solve this problem?

_____

5. c.   What must you know about the term "neutralization"?

_____

5. d.   How important is it to know what 2M, 1M, HCl, and NaOH mean and how they relate to each other?

_____

6.   Analyzing test answers: A simple question that might appear on a Biology test is "Define osmosis." Read the following two possible answers, and then answer questions a–d.

Answer A: Osmosis is the diffusion or movement of water molecules from a region of high concentration of water to a region of low concentration of water through a semipermeable membrane or plasma membrane.

Answer B: It is the movement of molacqules through a membrane.

6. a.   Which question received full credit?

_____

6. b.   Which answer repeats the word "osmosis"?

_____

6. c.   Which answer uses scientific terms?

_____

6. d.   Which answer has a word misspelled?

_____

CHAPTER 15

Analyzing Results of Tests and Assignments

Objectives

When you have read this chapter, you will be able to answer these questions:

1. What should I do with graded or corrected assignments?
2. How can I analyze my test results so that I can improve my test-taking skills?
3. How can I analyze my test results to detect how I should change the way I study?
4. What are the types of errors other students make on tests?

Written and math-based assignments are evaluated and given a grade. The thoroughness of the grading varies. If these graded assignments are a significant part of the final grade, they are marked more carefully. If they are a minor part of the final grade, the details of the assignments will not be evaluated, but the general quality of the work will be given a grade. Most objective tests are graded by machine. Fill-in, essay, and math-based tests are graded by the instructor or test evaluator. The comments, suggestions, and corrections recorded on each paper will vary from very few to many, from very general to very specific.

Written work and test answers are returned. If you **keep copies** of the **test,** you will be able to compare your results with the test at home. However, if you can't keep copies of the test, you will have to **arrange** to **see** a **copy** of the test to do this comparison. You have a right to do this and should do it. Make an appointment with your instructor. Don't moan and groan about the test to the instructor. A review of your mistakes is an important step in learning how to study, to complete future assignments, and to take tests. You want to learn from your mistakes and not continue to make the same kinds of mistakes. Remember, test-taking, assignment completion, and analyzing your results are the best ways to learn course material.

Returned Assignments

Keep all papers that are returned during the semester. Returned papers are your proof of completed assignments. If you plan to take another course from the same instructor, keep the papers for review. Papers graded in your present course will guide your work in future courses offered by the instructor.

If instructors discuss details of corrections and errors appearing on graded assignments, you should take notes on these comments. Turn to the question being discussed and record the notes on the returned assignment. If your work is correct, place exclamation marks next to the work. Since the instructor is spending time on this question, it might well be a *clue* that this type of information or problem will appear on a test.

If you have received a grade lower than expected and cannot figure out what you did wrong, you should make an appointment to see the instructor. Correcting errors and understanding why you made an error on an assignment will hopefully prevent you from making the same error on tests.

Analyzing Test Results

When a test is returned, you may feel satisfied with the results. If you are not satisfied, you must realize you have to do something to increase your level of satisfaction or to increase your grade. Obviously if you did not prepare for the test, you should not expect much satisfaction. The grade you receive on a test is a measure of your:

- knowledge and ability to apply that knowledge;
- comprehension and use of English;
- test-taking skills; and
- comfort and confidence level for taking tests.

Analyze all of your answers. Evaluate and compare correct and incorrect answers. Figure out why some were

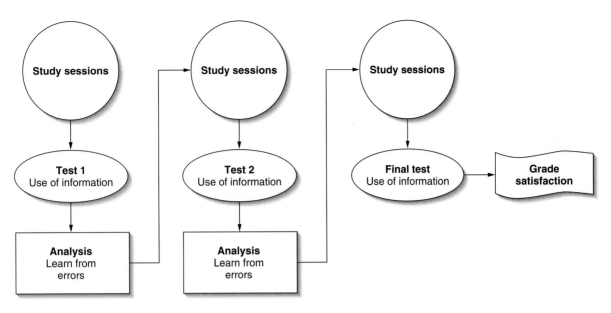

Figure 15.1 The analysis of test results should help direct your study for other tests. The analysis should also provide clues to help you improve your test-taking skills and reduce your test anxiety. The net result is your grades should improve.

correct and others were wrong. Regardless of the results, whether you got one or 20 answers wrong, analyze the questions you got wrong. Results on essay and math-based answers should likewise be analyzed. As you analyze the mistakes on tests, you set the stage to improve your study skills to enable you to learn, improve your comprehension and use of English, improve your testing skills, and reduce your test anxiety. In short, if you want to increase your grades on future tests, you must change your academic behavior (figure 15.1). To do this you should:

1. Identify what you have done correctly and incorrectly on tests.
2. Determine why you made the mistake(s).
3. Plan to change your study activities to prevent yourself from making the same kinds of mistakes on future tests. (Refer to chapters 3, 6, and 7 on study skills and time management).

What Is Wrong?

This is an easy question to answer for objective and fill-in tests! The answers that are marked wrong on the test identify your mistakes.

Detailed marking of essays or math problems by instructors will help identify what is wrong on subjective and math-based questions. Sometimes mistakes are not identified on graded tests. On an essay answer, you might only see a grade of 6 points out of 10 points. On a math problem, you might see an "X" or a " –10" next to the problem. Neither tells you what's wrong. You must figure out what's wrong by either redoing the answers to the questions or lis-

tening to the instructor's general comments and review of the answers to the questions.

Why Is It Wrong?

For each of the questions that is wrong, decide why it is wrong. Once you do this, you will be able to organize a plan of study to prevent yourself from making the same type of errors and to improve your academic skills and behavior.

There are many reasons why you have not answered questions correctly. Errors can be grouped into four categories:

- knowledge errors;
- English errors;
- lack of test-taking skills; and
- test anxiety.

Re-read the questions and answers that you have gotten wrong and decide which of the following categories and specific reasons accounts for your mistake.

Knowledge Errors

1. Did not learn because of lack of study.
2. Learned it, but did not retain the information.
3. Knew the information, but did not apply it properly.
4. Did not learn because it was not included in class notes.
5. Did not learn because you did not know about or do an assignment.
6. Did not read to reinforce the class notes.

7. Could not understand the figure.
8. Did not know the formula or equation.

English Errors

1. Could not comprehend sentence or question structure.
2. Test too long; reading too slow.
3. Vocabulary of nonscience terms not known.

Test-Taking Skills

1. Did not follow directions.
2. Did not define or focus on the test or qualifier terms in the question.
3. Changed the meaning of the question or thought the question was a trick question.
4. Skipped over part of the question to get to the answer.
5. Did not read or consider all choices.
6. Did not use time efficiently.
7. Changed right answers to wrong answers.
8. Recorded answers in the wrong spaces.
9. Did not see or read all of the questions.
10. "Read" too deeply into the question; did not accept question as it was written.

Test Anxiety

1. Nervousness caused a lack of concentration and gave you the "sweats."
2. Difficult questions caused a mental block.
3. Resorted to guessing just to finish the test.

Errors on Essays and Reports

Essays and reports are graded, but frequently the errors or deficiencies are not clearly identified. This places the analysis of why points were lost on the student. You must take the time to identify specific errors, plan to avoid repeating the errors, and possibly change your study techniques to help you respond more successfully.

To practice improving your essay-writing skills, rewrite essays that have been returned. Ask your instructor to review and evaluate your revision. Another way to help improve your essay-answering skill is to compare your answer to an answer written by a more successful student. Be sure to check for errors in knowledge or grammar.

As you analyze for errors keep the following list of errors in mind:

1. Did not know the meaning of key terms in question.
2. Could not recall information to answer question.
3. Information recalled was incorrect information.
4. Did not interconnect information in a meaningful sequence.
5. Did not include key words or other scientific terms in the answer.

6. Did not organize the essay with a thesis sentence, a series of paragraphs in the body, and a concluding sentence.
7. Did not number and answer all parts of the question asked.
8. Did not include figure when requested.
9. Used poor grammar and sentence structure.

Laboratory reports include more than essays. There is a particular format to follow as well as the proper construction of figures, graphs, and tables. Common errors on laboratory reports include:

1. format of the report is incorrect;
2. content of introduction and analysis lacks proper scientific relationships;
3. construction of graph is inaccurate or incomplete; and
4. results of the experiment are wrong.

Errors on Math Problems

See chapter 13, "Evaluate the Solution," on page 98.

Plan to Change

Once you have identified the reasons for your mistakes, you can start to prescribe a change in your academic behavior. If the errors are clustered in the anxiety category, you have to seek help to overcome your anxiety. If you have reading and vocabulary problems, you must seek help from the Reading Center at your college. You might consider enrolling in a reading comprehension course.

If you have made many test-taking mistakes, review the content of chapters 13 and 14. After you complete the review, re-do Exercise 14, paying particular attention to the skills checklist. Also, refer to chapters 3, 6, and 7 for information and guidance on study skills and time management. Set aside a specific amount of time, get into a test-taking mode, and practice taking tests! Use the following sources for tests:

1. See if your instructor would provide old tests for practice.
2. If your textbook has end-of-chapter-tests, take them.
3. Make up your own tests one day, take them the next day.
4. Have members of your study group compose short tests; exchange these tests, and take them. Return them to members of the study group to correct.
5. Buy a study guide and use the questions in the guide as test questions.

If poor grades are a result of knowledge errors, you must take steps to increase your knowledge and retention ability. You must change the way you are studying. Establish a study schedule, identify what you must learn, plan how to learn the information, and then go about learning according to the plan. Begin to use different approaches to learning the information. Don't just read the book or just review your notes. Do the things suggested in various chapters of this book. Use a study group to assist and stimulate your learning.

Comparing Mistakes to Your Notebook

Compare the questions you got wrong to the content of your notebook and weekly summaries. Can you find the answers in the notebook and summaries? Try to judge whether your notes are complete. Refer to chapter 5 on Listening and Taking Notes. If information about the question you got wrong is missing from your notes, you must ask whether the instructor covered the material or whether your notes are complete or incomplete. Plan to improve your note-taking skills if that seems to be the problem. If you think the instructor did not cover the material, then meet with the instructor to discuss why you should have been able to answer that question. Did you study your notes enough? Did you forget what was in your notes? Did you enhance your notes with corrections and additions?

Comparing Mistakes to the Textbook

Try to determine where the content of the questions you got wrong can be found in the text. Had you read that part of the text? Was the information part of a reading assignment? Did you consider the information in the text too unimportant to study? Was the information in the question part of the headings or subheadings of the textbook? (Refer to chapter 8, Use of Textbooks.)

Comparing Test Questions to Your Questions

Remember it was suggested that you use a portion of the left page of your notebook to record questions you have created. See chaper 5 on Listening and Taking Notes. The last comparison you should do is between the instructor's test questions and your questions. How did the questions you recorded in your notebook compare to the instructor's test questions? Did the types of questions you asked yourself help or hinder you in studying for the test? What kinds of information did the instructor ask questions about: facts, concepts, or application of information?

Results of Analysis

After analysis of your marked test, you will have identified your errors and decided why you made the mistakes. You will have also learned about parts of the test with which you were most successful. You can begin to plan actions to further improve your academic behavior. These efforts will hopefully lead to better grades. You should also have identified whether your notes are complete or not, whether you are using the textbook efficiently, and whether or not you have studied effectively. Resolve to correct the deficiencies as much as you can or want to. Take better notes; review the text; prepare your own questions, read, review, and recall more frequently; and make up more effective summaries, information charts, and concept maps. Practice problem-solving and test-taking. In other words, study more efficiently and effectively. Remember, you must make efforts to connect the base of the learning pyramid to the top of the learning pyramid.

If, after your analysis, you can't decide or understand why you made the error, ask your instructor or science tutor for help and guidance.

Review

1. It is important to keep all assignments and tests; they are the receipt of your work.
2. To analyze your results of a test, you should identify what mistake you made, decide why you made the error, and plan a change in your academic behavior to avoid making the same errors in the future.
3. Grades measure your knowledge and your ability to apply your knowledge.
4. Grades are influenced by your test-taking skills and the level of your test anxiety.
5. After organizing a plan to reduce the errors you make on a test, you should increase what you know and retain; improve your ability to use your knowledge; alter and improve your test-taking skills; and reduce the levels of your test anxiety.
6. If you compare the question you got wrong to your class notes, to the readings you did, and to the questions you created, you should be able to take steps to improve your notes, increase your reading comprehension, and craft better study questions.

Name _____

Date _____

Analyzing Results of Tests and Assignments

. . . .

1. List the four things measured by tests.

_____

_____

_____

_____

2. What are the three components of analyzing the results on your tests?

_____

_____

_____

3. Review the list of reasons that account for the following types of errors. Check the kind of errors you tend to make on tests.

Knowledge Errors

____ a. Did not learn because of lack of study.

____ b. Learned it, but did not retain the information.

____ c. Knew the information, but did not apply it properly.

____ d. Did not learn because it was not included in class notes.

____ e. Did not learn because you did not know about or do an assignment.

____ f. Did not read to reinforce the class notes.

____ g. Could not understand the figure.

____ h. Did not know the formula or equation.

English Errors

____ a. Could not comprehend sentence or question structure.

____ b. Test too long; reading too slow.

____ c. Vocabulary of nonscience terms not known.

Test-Taking Skills

____ a. Did not follow directions.

____ b. Did not define or focus on the test or qualifier terms in the question.

____ c. Changed the meaning of the question or thought the question was a trick question.

____ d. Skipped over part of the question to get to the answer.

____ e. Did not read or consider all choices.

____ f. Did not use time efficiently.

____ g. Changed right answers to wrong answers.

____ h. Recorded answers in the wrong space.

____ i. Did not see or read all of the questions.

____ j. Read too deeply into the question; did not accept question as it was written.

Test Anxiety

____ a. Nervousness caused a lack of concentration and gave you the "sweats."

____ b. Difficult questions caused a mental block.

____ c. Resorted to question just to finish the test.

4. List the errors that are commonly made on essay tests.

_____

_____

_____

_____

_____

5. List the errors that are commonly made on math-based problems.

_____

_____

_____

_____

6. List four actions that you will take to reduce the numbers of test-taking errors on your next test.

_____

_____

_____

_____

7. List the actions you plan to take to increase the amount you learn and improve the retention of what you have learned.

_____

_____

_____

_____

_____

8. Check which of the following is true and can help increase your knowledge or improve your retention of knowledge.

_____ The glossary helps you define key words.

_____ Headings and subheadings will help identify important information in the text and manual.

_____ Headings and subheadings can be turned into sample test questions.

_____ Boldfaced or italicized words are important words to add to your vocabulary.

_____ Chapter summaries outline the chapters' content.

_____ Figures help you envision the discussion in the text and the experimentation in the lab. If you know how to explain the figures, then you will know much of the content.

_____ Sample solutions help you form the basis to solve problems at the end of the chapter.

_____ Treat end-of-chapter or end-of-experiment questions as test questions. They will help you direct your study. Concentrate on the questions that relate to the topics discussed in lecture or the experiments performed in the laboratory.

9. If you have had poor results on a test, check which of the following activities you should include to change your academic behavior.

_____ Try to keep things simple. Don't defeat yourself by studying everything. Study the topics the instructor covered.

_____ Establish specific objectives. Answer the _how, when, where, what,_ and _why_ questions you have generated.

_____ Break big tasks down into smaller ones.

_____ Practice solving problems.

_____ Study in a study group.

_____ Study frequently; repeat learning the material in different formats—notes, figures, summaries, charts, and maps.

_____ Use tutors to help you clarify information and confirm you are learning.

_____ Realize when you study you're doing something good; it is good for your academic health.

_____ Assume the responsibility to study. Remember, it is all your responsibility and for your benefit. **Excuses don't count.**

_____ Don't delay, schedule 6 to 12 hours each week to study science.

_____ Practice the study skills until they become second nature to you. Internalize the study skills.

_____ Motivate yourself to learn the subject material.

_____ Realize it is satisfying to become competent in an academic area.

Mnemonics

Mnemonics are short verbal devices that help you recall a series of facts. The verbal device is a code to the series of facts you must learn. You can make up your own mnemonics or learn established ones. Mnemonics can be silly sentences, jingles, or acronyms. (An acronym is a word formed from the initial letters or letters of each of the successive parts or major parts of a compound term or series of facts.)

Silly Sentence Mnemonics

Geology

"*C*amels *O*ften *S*it *D*own *C*arefully. *P*erhaps *T*heir *J*oints *C*reak. *P*ersistent *E*arly *O*iling *M*ight *P*revent *P*ermanent *R*heumatism."

The first letter of each word refers to the geological time periods:

Cambrian, Ordovician, Silurian, Devonian, Carboniferous, Permian, Triassic, Jurassic, Cretaceous, Paleocene, Eocene, Oligocene, Miocene, Pliocene, Pleistocene, Recent.

Anatomy

"*L*azy *F*rench *T*oads *S*it *N*aked *I*n *A*nticipation."

This refers to some of the nerves passing through the skull:

Lachrymal, frontal, trochlear, superior, nasal, inferior, and abducent.

Physics

"*V*ultures like it *A*ll *R*are."

In the study of electricity, Ohm's law states that *volts* are equal to *amperes* times *resistance* ($v = ar$).

Some people find mnemonics helpful; others find they just add to the things one must remember. Students working in study groups seem to use and share the humorous mnemonics more frequently. Maybe you have to hear them to believe them. You should definitely consider using mnemonics as a tool to help you learn material and prepare for tests.

Exercise Resource

This appendix gives hints and partial answers to selected chapter exercises. Not all exercises are included. You will not find the answers to any of the exercises in this appendix. Answers must come from you. Some answers can be found directly in the chapters of this book. Certain exercises are more challenging than others. For all of them, compare your answers with those of your classmates, and discuss your answers with your instructor. You will find that it helps to discuss the exercises, especially the more challenging ones.

These exercises are designed to encourage critical thinking, synthesis of information, practice of concepts, and awareness about your learning style and study skills. You will also be developing and verifying the skills of answering questions. Completing these exercises will take time, but it is time well spent. In the end, your confidence for learning science will be boosted.

Exercise 1

1. a, b. Climate for Learning—Self-Evaluation
   There are no right or wrong answers to this evaluation, just honest judgements on your part.
   c. Improving Your Climate for Learning
   Example answers:
   Condition 3: I have a home or dorm environment that makes studying comfortable and possible. Action:
   a. Since I can't study at home, I will study in the college library, and I will also use the local library.
   b. When I do try to study at home, I will try to make my family aware that my study time is my private time and that they should not disturb me.
2. This is a discussion question. First, you would define what you think *scientific knowledge* and *neutral* mean, then discuss these with an example or two. Read page 1.
3. You can find definitions of "qualitative" and "quantitative" in a good college dictionary.
4. Information to form this hypothesis is found in the figure's caption.
5. See "Meeting the Prerequisites" and figure 1.3.
6. b. Compare your essay to the discussion of the scientific method, page 2, 3.
   c. Refer to chapter 8 about Internet use.
7. a. Your answer will be a personal response; there are no right or wrong answers.
   b. Compare SUDS to definitions of quantitative and qualitative.
8. Discuss your answers with classmates and your instructor.

Exercise 2

2. See chapter 10 on analyzing figures. Describe the flowchart, the meaning of the key terms, and how they relate to each other.
3. The answer to the first part of this question is found in the figure legend. For the second part of the question, review your diagram with classmates and your instructor.
4. See chapter 2.
5. Here is an example answer:
   If my learning style clashes with the instructor's teaching style and attitude, I will: focus on listening, observing, and studying the information, and not worry about the instructor;
6. a, b. See chapter 2.
7. This list depends on personal standards and goals.
8. a, b. Ask your instructor if a learning center is on campus and whether study groups can meet there.
9, 10. Discuss your answers with classmates and your instructor.

Exercise 3

1. a. Responses will vary.
   c. This is a personal evaluation about which "hemisphere" you are inclined to favor. You might want to check out a book in the library that would provide more information about this topic.
   d. Discuss your answer with classmates and your instructor.
2. a. This is a personal evaluation. You should be able to discuss the components of the pyramid and agree or disagree with the amount of learning for each component based on your learning experiences.
   b. Discuss your answer with classmates and your instructor.
3. There are several potential answers. An example answer is: Study groups can help connect the base to the top of the learning pyramid.
4. There are several potential answers. An example answer is: Self-testing with questions is a way to monitor learning.
5. Answers to this matching question could be debated. This question is asked to stimulate thought about the process of learning.
6. Refer to chapter 3. Discuss your answers with classmates and your intructor.
7. a, b. You "x" the study skills or activities you routinely use, but do this honestly. You might review notes, but do you correct and enhance the notes? You read and

re-read a text, but do you have clear objectives for your reading? As you select new skills, remember you might not use these all the time, but realize they will always be in your "learning toolbox."

Exercise 4

1. Information to complete the Science Course Survey will come from the course syllabus or science classes.
    b. For every hour in classroom you should spend about two hours studying.
    d. Refer to chapter 2 for descriptions of different teaching styles.
    k. If not, you must use your class notes to guide your use of the textbook and/or software by investigating topics mentioned in class. If you have doubts about what should be studied in the text, ask your instructor.
2. See chapter about self-discipline and self-direction. Review answers with classmates.
3. The answer for this depends on your personal life and conditions.

Exercise 5

1. a, b.   This checklist is a chance to evaluate your notes and list actions to improve.
2. Depending on your note-taking skills, this could be a long list.
3. How will you adapt the skills outlined in this chapter to your note-taking style? Discuss your thoughts and answers with classmates and your instructor.
4. See chapter 5.
5. This relates to the results of the checklist. The list contains specific actions to improve note-taking skills. Are you willing to alter your note-taking style?
6. Comparing notes with someone will require self-direction. Will you do it?
7. Comparing your notes with the textbook will help you (a) identify the style of your instructor; (b) verify the important things to study; and (c) realize the similarities and differences of the class and textbook.
8. For example, recording questions your instructor asks might provide hints about the questions that will be on tests.
9. a, b.   Your instructor will answer this.
10. Did you check the spelling of the scientific terms? Are the questions you asked complete thoughts? Here are two sample questions about the sample page found in figure 8.1 of chapter 8.
    a. Define and give three examples of polymers and monomers.
    b. Compare the chemical structure of a polymer and monomer. (**Note:** This question not only asks you to think about this page, but you also have to link these terms to earlier discussions on the types of atoms found in molecules.)

Exercise 6

2. a, b.   Use your "Two D's" to complete a Weekly Time Schedule.
3. a.   Be sure to analyze your use of time. This is an exercise that is quantitative in nature. You'll be coming up with numbers of hours rather than saying "I studied for a while."
3. b-6,   8   Review your answers with classmates and your instructor.
7. The study skill activities are listed in chapter 3. The time each person takes to do a task varies. Revising your notes might take an hour, but could take as much as two hours.

Exercise 7

2. For example, you can start by first identifying key terms in the lecture figure. Then look up the key terms in the index to discover the pages that illustrate these key terms. Compare those pages to the lecture figure.
3. It is important to fill in these lecture gaps. Your instructor is one resource. What are others?
4. Refer to chapter 3 after you make an honest effort to answer this question.
    a. If you could not list many study activities or skills, then you demonstrated the fact that we all forget information we learned.
    c. A study activity card is a flash card to help you learn what to do.
5. If reading assignments are not given, you must assume the responsibility to direct your outside study. The index and the table of contents of your book will give you pages and chapters that relate to your notes. (You must have good notes to help direct your use of the textbook.) Use S + Q + 3R + P (chapter 8) to help you use the textbook.
6. It is important to rephrase the key terms and concepts into questions. This self-directed skill will help you use the information. There are also other reasons. Can you identify them?
7. Refer back to your answers to question 3 in exercises at the end of chapter 1.
8. Compare your answer to suggestions given in chapter 14. Use your answer to help prepare for an exam.
9. See chapters 9 and 10 for help on constructing a diagram.
10. The answer to this is achieved by following the outlining directions. The textbook page in figure 8.1 would be as follows:
    1. Polymers
        a. Macromolecule and polymers
            1. Examples (see figure 8.1)
        b. Monomers
            1. Examples (see figure 8.1)
    2. Making and breaking macromolecule
        a. Dehydration synthesis
            1. Hydroxyl and hydrogen

b. Hydrolysis (see figure 8.1)
c. Catalysis
d. Enzyme

11. Review your answers with your classmates and instructor.

Exercise 8

1. Survey you textbook by flipping through it. Complete the checklist.

2–7. See chapter and discuss your answers with your classmates and instructor.

8. One example of a learning objective is: learn the steps of $S + Q + 3R + P$.

9. This is a personal evaluation, but good students establish objectives.

10. a. $S + Q + 3R + P$ means Survey, Question, Read, Recite, Review, and Practice.

    b. Answers can be found in each relevant chapter section. For example: The term "Survey" means to become familiar with the material and create a term inventory.

11. The answer to this question results from the analysis of your present reading technique.

12. Hints for $S + Q + 3R + P$ of figure 8.1:

    Survey: List the 14 boldface terms.
    Question: Develop questions related to reading objectives. See chapter sections on the $S + Q + 3R + P$ technique.
    Read: Twenty other terms were mentioned in the reading. List these terms. You should understand each of these terms. Yes, writing in science is information dense. If you do not know some of these terms, you should use the glossary and the index to find their meanings. Flash cards could be made for all new terms. The terms you would definitely have to know would be the terms mentioned in lecture or lab about this section of the book.
    Recite and Review:
        How long would it take you to recite and review the information about all these terms?
    Table 2.2
    The information in table 2.2 gives examples of macromolecules (polymers) and their respective subunits (monomers).
    Figure 2.16
    List the symbols and terms in caption and body.

The figure symbolizes what has been described in the paragraphs. A comparison of the figure and the paragraphs during the review would help you understand the information and achieve the two objectives for reading this page.

A concept map or outline could be created to help you review and practice using the information.

13. Will you include $S + Q + 3R + P$ in your "learning toolbox"?

14. a. Twenty-seven to 35 terms might be listed.

b. Estimate how long it would take to make one flash card and learn the term or phrase on that card. Multiply by the number of flash cards.

15, 16. See chapter.

17. For example, CD-ROM information can be used to check the accuracy of notes.

18. This question stresses the importance of defining a search problem and a search objective and identifying search key terms. If you had trouble with this search, then seek help from a friend or assistants at a computer center. This exercise should teach some pitfalls of research on the Internet and help you to search more effectively if you choose to use this method of research.

Exercise 9

1–7. See chapter information and discuss your answers with classmates and your instructor.

8. You might not know the names of the molecules in this chemical reaction. Remember, if a chemical reaction like this is discussed in class, you must learn the names of all of the elements and the names of the molecules! All of the symbols have an "English" meaning. Check with your textbook, instructor, and classmates.

9. a, b. Use the information from the answer to question 8.
    c–k. Review your answers with your classmates or instructor.

10, 11. Review and discuss your answers with your classmates or instructor.

Exercise 10

1. This checklist attempts to make you aware of your "figure literacy."

2. Refer to figure analysis tips in this chapter.

3. Think of $S + Q + 3R$ and compare your recall of the figure analysis technique to the "Figure analysis study card" you created.

4, 5. Review your answers with your classmates or instructor. All answers can be found within the figures.

6. a–e. Review your answers with your classmates and your instructor. All answers are found within the figure.
    f. See "Graphs" section in this chapter and compare to the graph in this question.

7, 8. Refer to chapter 10 and figure 10.2 and use a similar method to illustrate concepts. Review your answer with your classmates and your instructor.

Exercise 11

11. a–e, h, i. Most of the answers are in the questions and diagrams. While these questions may seem challenging and a lot of work, your time and effort will be worth the work. These questions will help you to learn how to review and analyze figures and synthesize information.

Review your answers with your classmates, a study group, and your instructor. You will be a better problem-solver if you take the time to complete these exercises and review them with others.

Answers 11f

When graphing this data, you should follow the format for graphing given in chapter 10. Check your answer with your instructor.

Answers 11g

Refer to chapter 10 and figure 10.2 to learn about creating figures. Use these resources to help you to draw diagrams for this question. Review your answers with your classmates and your instructor.

Exercise 12

1. This is a personal inventory of your study habits.
2. A group discussion should help you focus on the characteristics of good academic behavior.
3. The list of study activities and skills is given in chapters 3 and 7.
4. This essay should be made of three parts: the introductory sentence or paragraph, the body, and the concluding sentence or paragraph. The essay would begin with "The format for a formal laboratory report includes:" You would list the parts described in chapter 12. The body would describe the contents of each of the parts listed. The concluding statement would essentially repeat the introductory statement.
5–7. Refer to chapter 12. Review your answers with your classmates and your instructor.

Exercise 13

1. This is a personal inventory to allow you to determine your problem-solving characteristics.
2. This question allows you to evaluate the characteristics of a good problem-solver and invites you to assess yourself.
3. See figure 13.1.
4. The essay should be composed from either of the outlines. After you have written the essay, ask your instructor or the writing center to evaluate the essay.
5. In question (2) there are 7 key words. In question (3), four key words. In question (4), 6 key words.

6. Review your answers with classmates and your instructor.
7. a. For example: A triglyceride is synthesized by . . .
8–16. Review your answers with classmates, within a study group, and with your instructor. These questions will help you to practice skills for answering essay and math-based questions in science.
12. f. For this question, there are the same number of equations as there are parts to the question.
15. a. There are 8 key terms in this question, can you list all of them?

Exercise 14

1. This is a personal checklist. Realize these characteristics relate to study skills, test-taking skills, test anxiety, and time management.
2. This checklist applies mostly to objective tests, but parts can be applied to essay and math questions.
3. Compare the key words you have underlined with someone else's. Note that you should underline the test or qualifier terms and the scientific terms.
4. a. b. Refer to the chapter and the S + Q + 3R technique.
   c. There are 7 key words.
   d. For example, "Name" (list) is a test/qualifier term.
   e. The second part of the question would be answered with two paragraphs: (1) A paragraph describing the colors in the acid pH ranges; and (2) a paragraph describing the colors in the base pH ranges.
5, 6. Refer to this chapter. Review your answers with your classmates and instructor.

Exercise 15

1, 2. Refer to this chapter. Review your answers with your classmates and instructor.
3–5. These questions help identify the type of errors you make on assignments and tests. Knowing this, you should take action to change.
6–7. What changes will you make? See pages 110–113 and questions 8 and 9.
8–9. These questions contain choices of actions that relate to study skills, test-taking skills, and science anxiety.

B I B L I O G R A P H Y

Ambron, J. "Writing to Improve Learning in Biology." *J.C.S.T.,* February 1987, 263–66.

Annis, L. F. *Study Techniques.* Dubuque, IA: Wm. C. Brown, 1983.

Bogue, C. *Studying the Content Areas.* Clearwater, FL: H & H Publishing, 1988.

Bosworth, S., and M. A. Brisk. *Learning Skills for the Science Student.* Clearwater, FL: H & H Publishing, 1986.

Briscoe, C., and S. U. LaMaster. "Meaningful Learning in College Biology Through Concept Mapping." *The American Biology Teacher.* Vol. 53, No 4. 214–18, April 1991.

Carver, J. B. "Ideas of Practice: Plan-Making: Taking Effective Control of Study Habits." *Journal of Developmental Education,* November 1988, 5(2), 26–29.

Cooke, L. M. "Design for Excellence. How to Study Smartly." National Action Council for Minorities in Engineering, Inc., New York.

Crafts, K., and B. Hauther. *Surviving the Undergraduate Jungle.* New York: Grove Press, 1976.

Davis, A., and E. G. Clark. *T-Notes and Other Study Skills.* Metamora, IL: Davis & Clark Publishing, 1985.

Elliott, H. C. *The Effective Student.* New York: Harper & Row, 1966.

Farrar, R. T. "College 101." *Petersons Guides.* Princeton, NJ, 1984.

Foo, S. *Noteworthy Success.* Cornell Countryman, Vol. 36, p. 11, October 1988.

Fry, R. W. *How to Study.* Hawthorne, NJ: The Career Press, 1989.

Geoffrion, S. *Get Smart Fast: A Handbook for Academic Success.* Saratoga, CA: R. & E. Publishers, 1986.

Glaser, William. *The Quality School: Managing Students without Coercion.* New York, Harper Collins, 1992.

Haburton, E. "Study Skills Packet." Valencia, Valencia Community College, CA, 1978.

Heiman, M., and J. Slomianko. *Methods of Inquiry.* Cambridge, MA: Learning Associates, 1986.

Kay, R. S., and R. A. Terry, eds. *How to Stay in College.* Washington, D.C.: University Press of America, Inc., 1978.

Kesselman-Turkel, J., and F. Peterson. *Study Smarts. How to Learn More in Less Time.* Chicago: Contemporary Books, Inc., 1981.

Kesselman-Turkel, J., and F. Peterson. *Note-taking Made Easy.* Chicago: Contemporary Books, Inc., 1982.

Knowles, M. *Self-directed Learning.* New York: Association Press, 1975.

Langan, J. *Ten Steps to Improving Reading Skills.* Cherry Hill, NJ: Townsend Press, 1988.

Maiorana, V. P. *How to Learn and Study in College.* Englewood Cliffs, NJ: Prentice-Hall, Inc., 1980.

Mallow, Jeffery. *Science Anxiety.* Clearwater, FL: H & H Publishing, 1977.

Maxwell, Martha. *Improving Student Learning Skills.* San Francisco: Jossey-Bass Publishers, 1981.

McMillan, Victoria E. *Writing Papers in the Biological Sciences.* 3rd. ed. Boston, MA: Belford Books, 2001.

Miles, C., and J. Rauton. *Thinking Tools.* Clearwater, FL: H & H Publishing, 1985.

Ohm, H. *Note Taking and Report Writing.* Palo Alto, CA: California Education Plan Inc., 1989.

Strauss, M. J., and T. Fulwiler. "Interactive Writing and Learning Chemistry." *J.C.S.T.,* February 1987, 256–62.

Swartz, C., "The Teacher-Centered Lecture Method." *The Physics Teacher,* October 1995, 422.

Tonjes, M. J., and M. V. Zintz. *Teaching Reading, Thinking, Study Skills in Content Classrooms.* Dubuque, IA: Wm. C. Brown, 1981.

Trillin, A. S. and Associates. *Teaching Basic Skills in College.* San Francisco: Jossey-Bass, 1980.